IMPACT-FIRST PRODUCT TEAMS

Define Success.
Do Work That Matters.
Be Indispensable.

Matt LeMay

Impact-first Product Teams
Copyright © 2025 by Matthew LeMay. All Rights Reserved.

For information about this title contact the publisher:
Matthew LeMay
https://mattlemay.com
book@mattlemay.com

ISBN: 979-8-9917573-0-0 (Softcover)
      979-8-9917573-1-7 (eBook)

Cover design by Todd Goldstein
Illustrations by Christi Williford
Interior design and layout by 1106 Design

Thank you so much for taking the time to read *Impact-first Product Teams*.

Want to order a whole bunch of copies
Email book@mattlemay.com.

Want some help putting the ideas fror
organization? Email consulting@mat
about my consulting services in the "
of this book.

Want recommendations for a great restaurant in downtown New York City, an overlooked late '90s indie rock record, or a passable low-lactose cottage cheese alternative? Email matt@mattlemay.com.

Praise for *Impact-first Product Teams*

This book is a must-read for product teams navigating today's turbulent business landscape. I love the case studies, the insights, and the focus on actions that can be taken at the team level. But most of all, I love how fun this book was to read. Too many business books are painful to slog through, but Matt's book is as charming as it is useful.

> —Christina Wodtke
> Author of *Radical Focus* and *The Team that Managed Itself*

Product teams have been losing credibility in recent years. We've hidden behind dogma, avoided accountability for the quality of our decisions, and alienated the rest of the business, all while complaining about a lack of autonomy. But the business won't tolerate it anymore, nor should they. A reckoning is upon our craft, and if we don't evolve, we risk becoming irrelevant. It's time to focus on impact, not dogma. Matt is wise. Read his book and follow his advice.

> —Dave Wascha
> Former CPO/CPTO at MOO, Photobox, Moonpig, Zoopla

Any product leader or aspiring leader needs to read this book and apply its lessons. I was fortunate to have Matt join our PM planning session and preview some of his key tenets as he was finishing this book. Matt's message resonated with our team: The most effective PMs and PM teams must be commercially savvy, speak the language of the business, and 'move the needle.' Everything we do in PM must support this directive, or we should stop doing it.

> —David Nash
> Veteran B2B SaaS Chief Product Officer

Impact-first Product Teams *sets the standard for how every product team should function and has become my go-to recommendation for all product professionals I work with. Whether you're early in your career or a seasoned professional, this book is a must-read for those who want to create real impact and thrive in the ever-evolving world of product development.*

> —George Barlow
> Chief Product Officer & Product Design Consultant

Impact First Product Teams is an inspiring-yet-deeply-practical guide to what good product management actually looks like. Each chapter equips readers with an actionable question they can use to interrogate what they are delivering for the business and reframe their ways of working to confidently produce more value. This book is essential for anyone involved in the building of digital products—especially in this unpredictable season— because it bolsters us with tested, detailed guidance on how to do our absolute best work and grounds us in the compassionate, steady reminder that our jobs are only a facet of who we are.

     —Chelsea Bullock
     Senior Principal Product Manager at Atlassian & Former
     Product Executive for Early-Stage Startups

Too many Product Managers fixate on applying the 'correct' development methodologies instead of focusing on desired business outcomes. As a result, too many product teams have become stuck in the Low-impact Death Spiral of long hours spent shipping features that sink without trace. Matt's excellent new book resets the discussion and will ensure that every product manager (and every product team) is answering the question of 'what should my team build and why?' in the most impactful way possible for both their customers and their business.

     —Paul Jackson
     CEO, Streme

I've seen too many teams getting stuck trying to do product 'the right way' and losing sight of the big picture. They bounce from disconnected OKR to disconnected OKR each quarter but slowly lose the trust of the business. This book is a brilliant toolkit of questions—hard but important questions— that product teams should be asking themselves in order to rethink their work and revitalize their relationships with leaders.

     —Tom Dolan
     Head of Product - Which?, Product Coach, and former Head of
     Product Management at Government Digital Service

Table of Contents

Author's Foreword

⁞⁞⁞⁞▶

This is not the book I thought I would write.

But over the last few years of my work with product teams and organizations, I started to notice a pattern that genuinely surprised me:

The product teams that were most closely connected to the overall goals and objectives of the business were the most efficient, effective, aligned, and *happy* product teams I encountered.

And the product teams that were floating far away from the overall goals and objectives of the business—or in some cases actively resisting them? Many of them were anxious, burnt out, embattled, and exhausted.

Before I started working in product, most of my professional life was spent making music. Coming from this background, it seemed outright bizarre to me that "the business" would be anything other than a source of unwanted stress and unreasonable pressure.

But the more product teams I worked with, the more consistent this pattern seemed to be. And the more time I spent with these teams, the more I began to understand why.

To cut a medium-sized business book short: **Product teams that put the business impact of their work front-and-center are *working with*, not *against*, the existential reality of their situation.** They recognize that their teams exist for a reason, and that reason is not to "make the company more product-led" or to show everyone the "right way" to do product development. They understand that navigating commercial pressure is part of their job, even when it doesn't align with the "best practices" they've read about. (Best practices which, as we will discuss, emerged from companies that likely do not face comparable commercial pressure.)

These **impact-first product teams** are able to prioritize their work more effectively because they understand what they are prioritizing their work against. They are able to work more closely together because they understand that success can only be achieved through their shared efforts. And they are able to *step away from their work* at the end of the day, because they recognize that their job is just that: a job.

Being an impact-first product team does not require a company-level reorganization or a shift to a new operating model. But it does require a few subtle but meaningful shifts in how teams think about and execute their work; shifts that usually start with some of the powerful questions we'll walk through in this book.

So, yeah, here I am, a skeptical, creative writer penning the foreword to a book about *business impact*. It's not the book I thought I would write but, against all odds, I think it might just be the book that today's skeptical, creative product teams need. I hope you enjoy it.

—Matt LeMay, London, England, August 2024

"There is nothing quite so useless as doing with great efficiency something that should not be done at all."

—Peter Drucker, Managing for Business Effectiveness, 1963

INTRODUCTION

The Business Is *Your* Business

I n December of 2023, beloved tech giant Spotify announced that it was laying off approximately 17 percent of its workforce. In a public statement released at the time, CEO Daniel Ek described the following challenge at the heart of his decision:

"Today, we still have too many people dedicated to supporting work and even doing work around the work rather than contributing to opportunities with real impact."[1]

Ek's words sent a shiver down the spine of nearly everybody who has ever worked on a product team, myself very much included. For many of us, understanding the "real impact" of our work simply hasn't seemed like part of our job. Executives think through all that business impact stuff, then hand off lower-level objectives, strategies, or initiatives for product teams to work towards. Heck, in many cases, those executives just *tell us* what to build. To put it simply, **many product teams have been able to treat the business as somebody else's business.**

As Spotify's announcement vividly illustrates, this hands-off approach is no longer an option in a world of limited resources, unstable macroeconomic conditions, and ever-advancing rounds of layoffs and redundancies. Product teams now stand to bear the brunt of business-level outcomes—which

---

[1] Daniel Ek. 2023. "An Update on December 2023 Organizational Changes." Spotify. December 4, 2023. https://newsroom.spotify.com/2023-12-04/an-update-on-december-2023-organizational-changes/

means that product teams must deeply understand their business-level impact, even if they are not explicitly tasked with doing so.

Accepting this new reality has completely changed the way I approach my work as a product consultant and advisor. I'm often tasked with helping teams implement "best practices" or otherwise improve the *way* they build products. But at this point, **I am much more interested in helping product teams build the right things than I am in helping them build things the "right way."** Product teams can go through all the motions of "doing it right," fix every procedural and process-related issue, and *still* spin their wheels working on things that have no real impact on the business's big-picture goals.

The ramifications of this problem are ubiquitous and devastating. Teams and individuals burn themselves out working towards arbitrarily "urgent" deadlines, only to find themselves utterly dispensable when the company falls short of its overall goals. To make matters worse, teams and individuals often have no idea what the company's overall goals even *are*, leaving them to measure success by how late they're working, how happy their managers seem to be, or how closely their product development process mirrors what they've read about in books and blog posts.

This book captures the directional shifts I've seen brave product managers, teams, and organizations make to begin doing truly *impactful* work. It gets to the root cause of many of the problems facing product teams at large and small organizations alike, and suggests a different path forward. It is in no way a guarantee that your team or organization will achieve its most important goals. But it will help make sure that your team's efforts are actually aligned with those goals.

How Did We Get Here?

Martin Eriksson's classic article "What, Exactly, Is a Product Manager?"—the one that originated *that* Venn diagram[2]—states quite clearly: "Product

---

[2]Martin Eriksson's article introduced the 3-way "tech/business/UX" Venn Diagram that has become ubiquitous and, at times, misunderstood in product circles. There's a great video of Martin and me discussing it on Mind the Product here: https://www.mindtheproduct.com/interpreting-the-product-venn-diagram-with-matt-lemay-and-martin-eriksson/

Management is above all else a business function, focused on maximizing business value from a product." Nearly every popular history of product management goes back to the (obviously outdated language-wise) "brand man" role introduced by Procter & Gamble in the 1930s to oversee the holistic financial health of an individual line of consumer products. We can all agree, at least in theory, that product management has *something* to do with business.

And yet, many of the product teams I work with feel completely disconnected from the broader business. They have passionate, pitched debates about the "right way" to do product development, the "right amount" of discovery work to do, the "right operating model" for the organization to adopt. Many of these "right ways" take the form of well-documented "best practices" from organizations that have largely become synonymous with successful modern product development.

What's the problem here? For starters, most of the organizations associated with these "best practices" **are in fundamentally different commercial positions from** *nearly any company seeking to adopt those practices.* Spotify, for example, did not post a profitable quarter until 2018, well after a generalized version of its operating model became *de rigueur* among product organizations. Amazon's "reverse press release" method can be a helpful tool for product teams wanting to work backwards from customer and business outcomes. But the very idea of working backwards from a "press release" can be distracting for companies that are struggling to achieve a sustainable baseline of usage and revenue. (A struggle that Amazon itself has not faced for some time.)

To be clear, many of these technology behemoths have certainly been able to achieve "success" on their own terms. But their terms are likely going to be very different from *your* terms. There is always something to be learned from the ways that *any* other company has approached the complex and interrelated challenges that come with building products. But the widespread belief that a small handful of largely venture-backed and often-monopolistic companies have somehow figured out how *every single company should build products, regardless of that company's business and/ or funding model,* simply doesn't add up.

The proliferation of one-size-fits-all "best practices," of sanitized case studies from Silicon Valley darlings, of "best vs. the rest" narratives, has created an environment where just about everybody working within the real-world constraints of most companies' business and funding models will *never* feel like their companies are doing things "the right way." The time has come for us to recognize that every product team at every company must find their own way. And that starts by understanding what their particular business needs to succeed.

Why "Impact First"?

The thesis of this book is twofold:

First: In this day and age, product teams can no longer afford to be disconnected from the business-level impact of their work.

Second: Because business-level impact depends on complex market dynamics, teams cannot simply *choose* to be high impact. They can, however, choose to put **impact** *first* when deciding what goals to work towards, what work to prioritize, and how to work together.

The kind of business-level impact that really matters cannot be controlled. Market conditions can change, the competitive landscape can change, pandemics and wars can break out, executives can get into turf wars that send company strategy flying in opposite and incompatible directions. There is no *guarantee* that a team's efforts will result in success for their business.

It's not terribly surprising that many teams are put off by this combination of high stakes and low certainty. The biggest challenge facing many product teams is not that their high-impact efforts are falling short, but rather that **they are prioritizing work that has no chance of delivering meaningful business impact in the first place**. Many teams prefer to chase low-impact goals that provide them with a sense of control and autonomy. Some such teams will *insist* that this is the "right way" of doing things; that goals *must* be atomized and cascaded until every team has its own tidy and self-contained purview. But when the business finds itself faltering, all of these teams will discover that "but we were doing things the way we're supposed to!" is a cold comfort.

So what does it mean to be an impact-first product team? **Simply put, choosing to be an impact-first product team means recognizing that our responsibility is ultimately to contribute towards a successful business.** And that questions of how we work to deliver on that responsibility are ultimately less important than the responsibility itself. It means that questions of whether we are "product-led" or "sales-led" are always secondary to the much more important question: "What are we doing to contribute to the success of our business?"

Why "Product Teams"?

This book is focused on the team level because the team level is where product work comes to life. The success of a product team is realized in that *team*'s work, not in the work of its individual contributors.

This book is also focused on the team level because, as we will discuss, it is ultimately incumbent upon product teams themselves to draw a straight line between their day-to-day work and the overall success of the business. This book is focused on the things that product teams can *actually do* to put business impact at the heart of their work regardless of the broader contours of their organization. Questions of impact-first executive leadership, company-wide operating models, and top-down incentives are for the next book.

A Line Through the Middle

I recently explained the concept of this book to a friend of mine who happens to be an excellent product manager. I was struggling to articulate how an "impact-first" approach to product complements the excellent writing and thinking that currently exists around things like strategy and discovery.

"I think I see it," she said. "What you're talking about is drawing a straight line from the stuff that teams do every day to the high-level business impact of that work. Most of the stuff that's written about product development is about the middle. But you're trying to make a direct connection between the most business-y stuff and the stuff that's just kinda . . . vibes." I laughed. I sighed. I was relieved. Yup, that's it.

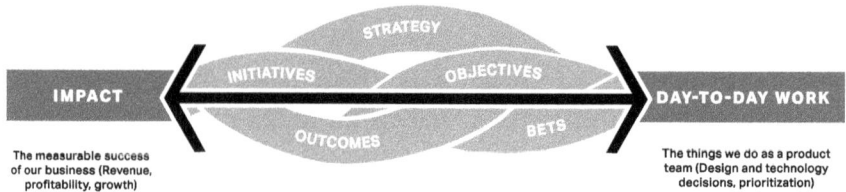

Figure i-1: A direct line between impact and day-to-day work

There's a good reason why most books and articles about product management focus on "the middle." The middle is where our impact-level ambitions turn into actionable plans. The middle is where we learn whether or not the things we *hope* will deliver business-level impact will actually matter to our customers.

The goal of this book is not to eliminate the middle. We *need* the middle, and there are countless books and articles that are helpful for making sense of the middle. **But if we exclusively focus on doing the middle "the right way," we still run the risk of putting a lot of time and effort into work that ultimately doesn't matter.** To make matters worse, if we overcomplicate the middle, we run the risk of it further separating us from the actual business-level impact of our work.

By and large, this book does not get into much detail about the middle. There are *great* books out there about strategy, discovery, initiatives, and aligning teams around customer outcomes. There are lots of different ways to approach the middle, depending on your organization, industry, and business model. This book is intended to complement any and all of those approaches by reminding us to ask the simple (but not always easy) question: *does all of this add up to a successful business at the end of the day?*

Goal-setting Beyond Frameworks

In keeping with the "line through the middle" theme, this book is not about any particular goal-setting framework. You can have your OKRs, your V2MOMs (a fixture of the extended Salesforce cinematic universe), or any emergent frameworks or approaches your organization has used to set goals. The ideas in this book are intended to help you **keep business impact front and center as you go about setting and executing against** your goals, regardless of the particular process or framework you use to create those goals.

Powerful Questions

Each chapter of this book contains a powerful question that teams can use to put—and keep—impact first as they go about the work of building products. In my experience, the biggest challenges facing most teams do not arise from answering these questions imprecisely, or even incorrectly, but from failing to ask them at all. These questions are intended to be discussed openly and fearlessly among the entire product team. If you'd like to bring these questions directly to your team, you can find them—along with other templates and resources from this book—at https://mattlemay.com/impact/resources.

First-hand Stories

By and large, the stories in this book are stories from my own experience, or stories told first-hand by other product and business folks. **Details have been changed to protect the innocent, and every specific number included in these stories is entirely made up.**

There are a lot of overused case studies from "best in class" companies out there, and they all look *very* different when told by people on the inside. My goal is not to present idealized "just do these things and your team will magically be better" stories, but rather to provide concrete examples of steps that folks have taken, what's worked, what hasn't worked, and what's been surprising.

Generally speaking, my goal here is not to present an idealized version of *anything*, but rather to share some ways that I've seen product teams proactively manage the conversation around business impact. Some may work in your context. Some may not. Trust your instincts, be brave, and remember that product work is always challenging work. And most of us wouldn't have it any other way. Sound good? Let's go!

CHAPTER 1

The High Cost of Low-impact Teams

▄▄▄▄▄▄➤

Early in my career, I found myself running one of two product teams at a midsized technology company. My team was responsible for the free, consumer-facing product, while the other team looked after the revenue-generating enterprise side of the business. This company had grown up around its consumer product, and the teams were resourced accordingly: My team had about seven full-time engineers and designers building cool new features, while the enterprise team had one part-time engineer maintaining an entirely separate code base.

I was a young, ambitious product manager, eager to show company leadership that I knew how to run a high-performing product team. I put together an impressive roadmap, full of grand plans for cutting-edge features and thoughtful reimaginings of our existing functionality. My team started having daily stand-up meetings and documenting our day-to-day work in an accessible, transparent backlog. We ran A/B tests, interviewed users, and monitored the success of new features we shipped. We were, by all accounts, doing product "the right way."

There was only one problem: none of the things we were doing "the right way" actually mattered. While I had been hyperfocused on building a "high-performing product team," I hadn't stopped to consider that our consumer product had no solid business model behind it. We were excitedly jumping from defensible feature to defensible feature, with absolutely zero sense of how these features were contributing to the overall health of the business, or offsetting the sizable expense of employing a seven-person product team.

Meanwhile, our barely-resourced enterprise team had been slowly but surely generating a truly impressive amount of revenue. They had pieced together a product that people were willing to pay for, and they were working closely with our sales team to find a steady stream of people who were willing to pay for it. Rather than shipping tons of exciting new features, our enterprise team was laser focused on turning our core product offering into a profitable business.

For all of my razzle dazzle, it was becoming clear that the best and only path forward for the business was to build a sustainable enterprise product, not a growth-focused consumer product. Believe me when I say that I fought this tooth and nail. I made grand promises about the incredible things my team would be delivering in the near future. I *insisted* that the scale of our ambitions couldn't be met with a humble, sales-driven enterprise product. I even complained about the enterprise team's obviously not-best-practice-following ways of working to anybody who would listen. But none of it mattered. We were costing the business money, and the enterprise team was making the business money.

After some very contentious meetings with the enterprise product management and company leadership, I grudgingly agreed to a plan: five of my seven engineers and designers would move to the enterprise team, and I would manage a "lean" team of one engineer and designer. In other words, I would watch my carefully built empire crumble before my own eyes and be relegated to obscurity while my tiny team did tiny things.

But that's not what happened.

Armed with more resources, the enterprise side of the business was able to focus on testing out new revenue-generating features while improving the stability of the core product. And my tiny team? We started rebuilding the consumer side of the product to better integrate with the enterprise side, with the goal of moving towards a single "freemium" model that would be easier to maintain. The higher-impact team had the resources it needed to steadily grow our revenue. And my tiny team was able to move quickly, efficiently, and nimbly to fit the changing shape of the business, even if we had to give up some of those "high-performing product team" rituals that we had once cherished. Best of all, the business was set on a path to profitability, which it achieved a few short years later.

To be clear, this was *not* an easy transition to make, nor was it a transition that I executed with much grace or patience. I feared that doing the right thing for the business would mean diminishing my own power and stature in the organization. And acting on that fear burnt up a good deal of valuable time, and nearly all the goodwill I had worked so hard to build up with my team. It certainly took me long enough, but I learned the hard way that realigning around high-impact work is much easier said than done.

In my case, as with many others, **the first step towards becoming a high-impact team was understanding that we were a low-impact team**. In this chapter, we'll look at some of the common signs of low-impact teams, and discuss why it is more important than ever for these teams to reevaluate their priorities with speed and urgency.

The Low-impact Death Spiral

For well-resourced organizations, the immediate costs of a low-impact product team are easy enough to manage. Sure, those well-compensated engineers, designers, and product managers might not be producing much of a return for the business, but their salaries are basically a rounding error, or a fraction of the office snack budget. Managers and directors are happy to have more folks reporting to them. And people working on low-impact teams can be excellent at their individual jobs, elevating the craft and capabilities of those around them. Some of the best experiences I've had in my career have involved working with high-performing people on low-impact teams, building fun things and solving interesting technical challenges without having to worry much about the health of the business overall.

Over time, though, the consequences of low-impact teams become less obvious and more entrenched. As these teams pump out frivolous new functionalities and meaningless tweaks, these bits adhere to the commercial core of the product, creating a thicker and thicker shell of shiny miscellany. This shell makes it even harder to enhance the product's core, while driving new features and functionalities even farther away from the *purpose* of that core. (Imagine three different teams adding bejeweled ornaments to the hood of a car, only to find it impossible to lift that hood and access the engine.)

Over time, this invisible and insidious dynamic becomes a self-reinforcing flywheel of doom that I call "the low-impact death spiral."

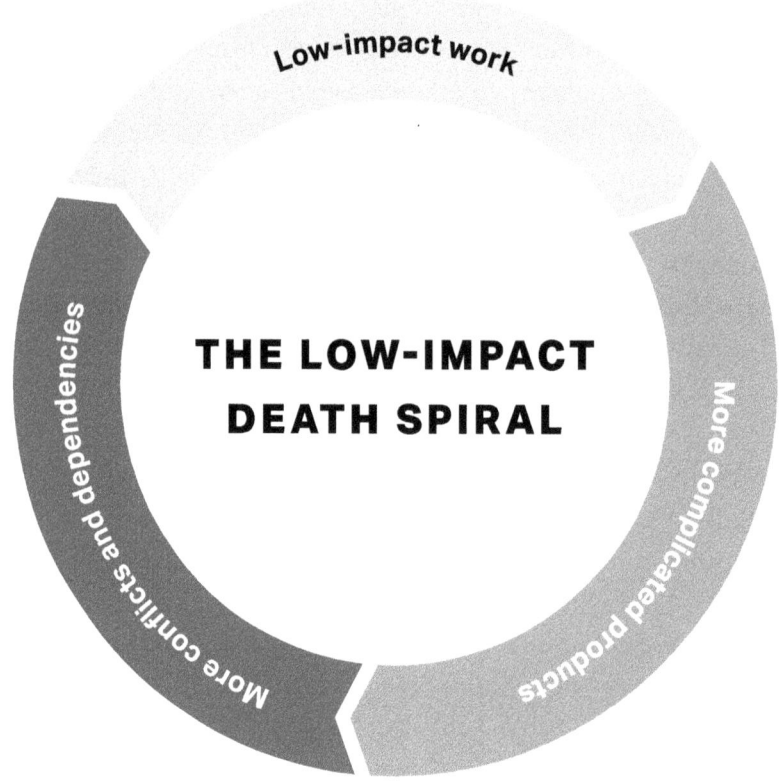

Figure 1-1: The Low-impact Death Spiral

Low-impact work creates more complicated products which, in turn, lead to more dependencies and conflicts to manage. Those dependencies and conflicts discourage teams from taking on work that touches on the product's commercial core. Which, in turn, encourages more low-impact work. Nearly every medium-to-large organization I've worked with is in the throes of this spiral, and most early-stage startups I've worked with are well on their way. If you've ever used a product that feels like a grab bag of incoherent features jammed together into a fragmented experience, that's the Low-impact Death Spiral working its evil magic. And if you've ever worked on a product team, you've probably contributed to it despite—or, perhaps, because of—your best efforts.

Imagine, for example, that you are a member of a product team working on a popular video conferencing application. The core areas of the product—sign-up, login, video, and text chat—are all built on a tenuous pile of old and new systems. Several teams all claim some degree of ownership over these areas, and executives are always a little bit twitchy about any major changes being made. So your team decides to give itself a break and work on something new: a "schedule a meeting with me" page that users can share with their contacts. You can build this without having to change much existing functionality, and competing products seem to be doing quite well. At the very least, you'll be doing no harm and you'll be left alone.

In only a few weeks, the feature launches. It's hidden in a relatively obscure settings page, so as not to cause too much trouble. It sees modest uptake from a few hundred existing users, and a handful of excited emails are sent through to customer support. Your team celebrates a job well done. Amidst all the logjams and bottlenecks of a complex organization, you managed to get something out the door quickly. A week later, the company's CPO cites your feature as proof that your product organization is capable of shipping new features with speed and efficiency. You did it!

A month later, though, things are not looking so rosy. Your company's revenue team—who you were able to avoid while launching your unobtrusive free feature—has been explicitly tasked with increasing the paid user base. They've been exploring enhanced meeting scheduling as a possible premium feature, and they're concerned that your team's work might negatively impact revenue potential if too many folks begin using it for free. Your team, meanwhile, has been in close contact with some of the feature's early adopters, and are devastated at the thought of hiding it behind a paywall.

In the ensuing weeks, multiple meetings are held to discuss the fate of your feature. These meetings are . . . not fun. Suddenly, the very executives who had celebrated your feature a month ago are peppering you with pointed questions and concerns. Your little feature, which initially seemed like such a quick and easy win, seems to come at an ever-growing cost of time and morale. As these costs become more visible, other product teams begin walking back any plans that might get caught up in the revenue team's inquisition. You hoped you could quickly push something into the product without having to touch its commercial core, but in doing so, you made that commercial core appear even more complicated and unapproachable. And as an added bonus, you managed to get the company's revenue team playing defense with your cute little feature rather than working to identify broader opportunities to deliver success across the organization.

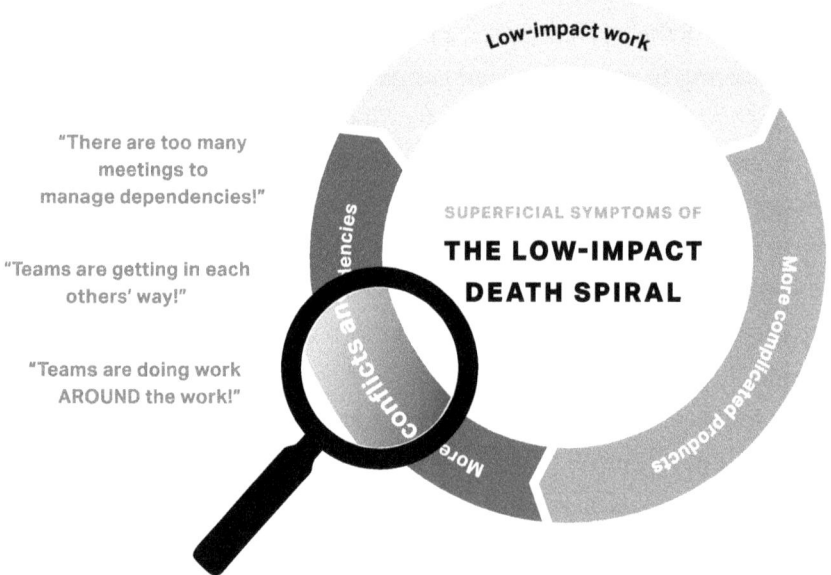

Figure 1-2: Superficial symptoms of the Low-impact Death Spiral

Having been through many such experiences, I can say with confidence that the organization-wide takeaway is rarely "teams should not be doing low-impact work that makes the product more complicated without delivering value for the business." It is, much more often, "we're having too many meetings to coordinate across teams" or, "teams are getting in each others' way" or, "we have too many teams doing the work around the work."

Ironically, "treating" the obvious symptoms of low-impact teams by adding more layers and processes to cross-team coordination often

makes the root cause even worse. As the administrative burden to take on high-impact work grows, teams are more likely to seek low-impact work. This, in turn, adds more fiddly and disconnected bits to the product. On and on it goes, until the next round of layoffs.

What does this mean for product *teams*? In short, it means that *nobody is coming to save you.* If you want to be an impact-first team, you can't wait around for the next reorg or leadership turnover. While executive support and organization-wide initiatives can certainly be helpful, **individual product teams *do* have the power to break the death spiral by stepping away from low-impact work.**

Impact-first Teams vs. Low-impact Teams

Though every team is different, there are consistent patterns that differentiate impact-first teams from low-impact teams. Of course, every team is shaped by the broader culture and habits of its organization. My goal here is not to cast aspersions on any teams that fail to live up to an abstract ideal, but rather to draw out some real-world patterns that might help teams understand the general contours of the challenges they're facing. And, in the coming chapters, how best to address those challenges.

⟶ **Impact-first teams fight to work on the most important things for the business.**

⟵ **Low-impact teams avoid this work if it requires too much coordination or invites too much scrutiny.**

The single most important characteristic of an impact-first team is that they actively seek out the work that they believe will most meaningfully contribute to the success of the business. This is often much easier said than done, especially since the most valuable work often invites the most scrutiny from the most stakeholders. One product team I worked with referred derisively to the executive "eye of Sauron" casting its menacing stare at any team brave and foolish enough to take on mission-critical work.

Impact-first teams accept scrutiny from the organization at large as a sign that the work they're doing is consequential to the organization at large. They recognize that the probing questions they receive from important stakeholders might help them better understand the broader context of the work at hand. They are open to having challenging conversations to coordinate with other teams if that coordination delivers better results for the business. And, above all else, they recognize that their job is ultimately to deliver impactful work, even when it is not *easy* work.

Low-impact teams are happy to plug away on their own little corner of the product, where they are safe from conflicts with other teams and difficult questions from senior stakeholders. And though their work might *feel* inconsequential to the broader organization, the features and functionalities they deliver only reinforce the Low-Impact Death Spiral.

Some low-impact signs to watch out for:

- Teams whose work is scoped to a particular "product area" or "feature set" without any business-level goals
- Teams that have zero regular points of connection with any other product teams
- Teams that dismiss potentially impactful work as too difficult to coordinate with other teams, or outside of their immediate purview as a team

⟶ **Impact-first teams commit to big goals that are largely outside of their control.**

⟵ **Low-impact teams commit to small goals that are fully within their control.**

As we will discuss in the next chapter, committing to impact-level goals means committing to goals that fall outside of our control. Success in real-world markets requires an untameable melange of luck, discipline, focus, and ambition.

Impact-first product teams know that the success of the business is ultimately determined by complex and interrelated dynamics involving customers, markets, and the world at large. They set and articulate their specific ambitions fearlessly, and seek out the work they believe is most likely to get them there. They recognize that the failure of the business will affect them whether or not they set and achieve lower-level, controllable, and ultimately irrelevant goals.

Low-impact teams reject outright any goals which are not entirely within their control. These teams are low-ambition by design, setting themselves up to claim victory while contributing minimally to the overall success of their business.

Some low-impact signs to watch out for:

- Teams that are only accountable for operational goals like velocity or number of features delivered
- Teams that reverse-engineer their goals from the work they already have planned
- Teams that broadly resist estimating impact because it's "too complicated" or "involves too many things outside of our control"

Impact-first teams keep their goals front and center throughout the planning and development process.

Low-impact teams only think about their goals during intermittent goal-setting exercises or "seasons."

I'm always baffled when I see entire organizations grinding to a halt for a quarterly goal-setting "season"—often referred to as "OKR season" for companies using the Objectives and Key Results framework—only to completely disregard those plans until the next scheduled season.

For impact-first teams, high-level goals are an ever-present factor in decision-making. As we will discuss in chapter 5, these teams are constantly checking their planned work against the impact they seek to drive for the overall business, even when doing so is neither obvious nor easy. These teams are always prepared to explain the impact of their work, even when they aren't asked directly.

Low-impact teams approach goal-setting "exercises" as little more than a frustrating interruption. They see setting goals and delivering products as two entirely different things, and the former can only get in the way of the latter. And, hey, nobody ever *asks* about those goals after "OKR season" is over, so why take it seriously in the first place?

Some low-impact signs to watch out for:

- Teams that only engage with goals during an isolated "season" or "exercise"

- Teams that go about setting the next quarter or year's goals without reflecting on the prior quarter or year's goals
- Teams where only one member (usually the product manager) regularly engages with higher-level goals

⟶ **Impact-first teams see product development processes as a means to an end.**

⟵ **Low-impact teams care more about doing things the "right way" than about doing the right things.**

The discourse around product development is full of "best practices," ranging from quick-and-dirty productivity hacks to complex organization-wide frameworks. These "best practices" provide ample fodder for teams that feel safer arguing about the right way to do things than actually working on the right things. And there are a *lot* of these teams.

Impact-first teams understand that tools and ways of working are simply a means to an end. These teams recognize that the primary takeaway from popular ways of working like "the Spotify model" is to *purpose-build your own ways of working* that help you achieve your own goals in your own specific context, not to "just do what famous company X does." They are constantly adjusting their ways of working to help them deliver results for their specific business.

Low-impact teams tend to conflate ways of working with the *results* of those ways of working, and to focus exclusively on the former. They devote their energy to finding the "right" roadmapping tools, rather than actually assembling a roadmap, let alone *building* the things on their roadmap. They have strong opinions about which frameworks should be used for setting goals, but they are reluctant to commit to any actual business-level goals that would help them make better day-to-day decisions.

Some low-impact signs to watch out for:

- Teams that seek or demand absolute, nonoverlapping role clarity as opposed to working together towards shared goals
- Teams that claim that they cannot do impactful work because their organizations are not "product-led" or "empowered"

- Teams whose goals and/or objectives are so complicated and framework-heavy that they cannot be easily repeated and require a deck or other document

Impact-first teams recognize that their own operation comes at a cost to the business.

Low-impact teams see any value they bring to the business as purely additive.

A well-staffed product team comes at a substantial material cost. As we discussed earlier in this chapter, some companies are better positioned than others to shoulder this cost, but at a certain point, somebody is bound to ask, "Why exactly is this team worth paying for?"

Impact-first teams recognize that their own work comes at a cost to the business, which means that the business is expecting *some* kind of return from that team. That might take the form of a financial return, a return in the investability of the company, or a return in the company's ability to project success in the broader market. Impact-first teams work tirelessly to understand what success means to the business at large, and how the business expects them to contribute to that success.

Low-impact teams see their own existence as an immutable and righteous truth of the universe. They celebrate anything and everything that could be seen as a success, even if that success is a drop in the proverbial bucket of their cost to the business. If and when they are asked to justify their value to the business, they react with outrage and indignation. And, unfortunately, by the time they're *asked* to justify their existence, it's usually too late.

Some low-impact signs to watch out for:

- Teams that avoid talking about revenue, value exchange, and/or the company's overall business model
- Teams that celebrate any minor movement in success metrics, even if these movements are trivial compared to the team's cost to the business
- Teams whose members measure their individual success by how hard they're working or how much they "own" rather than their contributions to the team and business

Even the most impactful product teams I've worked with fall into low-impact patterns from time to time. The challenge for real-world teams is to be aware enough of these patterns to resist their pull, break the Low-impact Death Spiral, and make meaningful contributions to the overall success of the business.

The Accountability Conundrum

At the heart of many low-impact teams is the same decidedly reasonable question: *"How can we be held accountable for things outside of our control?"*

The answer to that question is not terribly fun to acknowledge out loud, but it bears stating plainly: **the overall success of a product *is* outside of your control.**

You can do everything within your power to build a product the "right" way only for that product to fail. And if the product you work on *does* fail, your team will likely find itself facing real-world consequences even if you did your very best. Businesses ebb and flow, competitive landscapes shift unexpectedly, changes in customer needs happen slowly then all at once. There is simply no way to control the myriad interconnected variables that lead a product to success, no matter how many "best practices" you put in place.

Is it fair, then, to ask individual product teams to be accountable for making a contribution to the success of their product and business overall?

The answer here is, perhaps, even *less* fun to acknowledge out loud: **a product team *already does* bear the consequences of the success of their product and business overall.** In other words, product teams are *already* accountable for things outside of their control, whether or not they accept it.

If your team is working on something that does not deliver value to the business, then the business likely won't see the value of your team. And if your business goes *out* of business, then there are obvious ramifications for every individual on every team throughout the organization.

None of which is to say that the term "accountability" can't be weaponized. Simply expecting a product team to magically turn money into *more* money is likely not going to set that team up for success, nor leave them feeling terribly supported and understood. But if a product team participates in defining *how much* money they will seek to generate, why that money matters to the business, and how they intend to generate that money, then they are in a much stronger position to be seen as a successful part of the business overall.

The real question for product teams, then, is whether they are willing to play an active role in aligning their accountability with the success of the business overall. Time and time again, I've seen that the risk of *not* having a perspective on your team's impact starts to outweigh the risks and challenges of having an incomplete or imperfect perspective. As we will discuss in chapter 2, **conversations about impact can actually be a powerful lever for creating more clarity around goals, risks, and expectations.** But for teams that are used to seeing all questions of business impact as outside of their purview or beyond their control, these conversations can be quite challenging indeed.

? ONE POWERFUL QUESTION
"If you were in charge of the company, would you fully fund this team?"

It is not easy to get a team to acknowledge that it is low-impact. If you ask somebody what impact their product team is having on the business, you might receive a word salad peppered with "indispensable"s and "mission-critical"s. You might receive a defensive soliloquy about how the team's contributions are massive but underappreciated. You might receive an expression of pure outrage that you would even ask such a question, given that the organization itself has failed to present cohesive goals or a reasonable strategy. But what you likely *won't* receive is an honest "I have no idea how to answer this question, and the fact that you are asking it makes me extremely nervous."

For that reason, I like to start with a question that requires both a change in perspective and an unflinching look at the actual stakes involved: **"If you were in charge of the company, would you fully fund this team?"**

This question compels people to consider—often for the first time—that the very existence of their team comes at a cost to the business. For teams that have happily jumped between bits of defensible but low-impact work, this can be a major shift. If your team is costing the business hundreds of thousands of dollars a year, what are you doing to deliver a return on that investment?

Even a slight pause in answering this question can open up a critical space for individuals and teams to reevaluate the work they're doing. Twice in my career, asking this question has compelled a team to substantially shift its goals, or even disband itself altogether.

In one case, a team decided that its responsibilities for a single *feature area* had, in effect, limited its ability to deliver enough impact to justify its cost. They recognized that giving up their small and well-bounded territory would likely require more coordination with other teams (we'll discuss this more in chapter 4), so they chose instead to focus their team on the holistic needs of an entire user segment.

The other team—a group of highly skilled senior engineers and designers tasked with working on "connective" pieces of a complex marketing platform—made the even more dramatic decision to effectively disband their team. They then reallocated their valuable time and expertise to teams that were already working on the most impactful part of the platform such as new user onboarding. While the idea of putting together a team to address the "bits between the bits" had made sense in theory, it was proving too difficult in practice for this team to make a major difference without being more closely aligned with larger and longer-standing product teams.

These decisions were not taken lightly, nor were they easy or painless to put into practice. But in both cases, the individuals on the teams chose to align their success with the success of the business. And do so on their own terms, before they were forced to do it on somebody else's.

To Summarize . . .

In most organizations, being a low-impact team is much easier than being an impact-first team. Impactful work can invite executive scrutiny and require extensive coordination across teams. Low-impact work offers product teams a sense of control and certainty, all the while avoiding difficult conversations about the company's overall goals and ambitions.

In the longer term, though, things rarely work out well for low-impact teams. Low-impact work often makes things *worse* for the organization at large, creating a kind of "death spiral" where new features and functionalities create additional administrative burden without helping the business achieve its most important goals. And when teams can't speak directly to the value they are providing to the business, they run the risk of somebody else answering that question for them—and shouldering the consequences of that answer.

In the coming chapters, we'll discuss how to turn a low-impact team into an impact-first team, starting by defining what exactly "impact" means in the first place.

Uniting a Product Team Around a Single (Nonrevenue) Impact Metric

Randeep Sidhu
Product Director, GovTech product

In June of 2020, I was called up out of the blue to advise on the National Health Service's ongoing efforts to build a COVID-19 app. It wasn't going well. The first version of the app was shut down and the team disbanded after only three months. My background in health tech was in building products at scale, but in a way that still helped underserved communities. Some senior folks at the NHS figured I might have a valuable perspective on what they needed to do differently when rebuilding this app.

I didn't pull any punches on these calls, and insisted the NHS prioritize the groups that were being disproportionately hurt by the pandemic, those who are often overlooked. My belief, which has been proven out, is that building for the bottom of the pyramid helps everyone, so being inclusive is the way to have the biggest impact.

Surprisingly, those initial phone calls were followed days later by an offer to join the NHS full time as director of their COVID-19 app. I had to leave my previous employer, and three days later walked into an absolute maelstrom. I hadn't had time to prepare, we had no existing team, and we were in the middle of a pandemic.

The most immediate challenge we faced was simply working at the pace that the situation demanded. We needed to make decisions quickly, and we needed to do so in an environment of fast-changing and emergent information. The combination of pressure and uncertainty made it very hard to align stakeholders. And we had a lot of stakeholders, ranging from government agencies to privacy advocacy groups to literal heads of state (with Google and Apple thrown in too!)

It was clear from the outset that we wouldn't be able to move at the necessary speed unless we had a way to align all these stakeholders around an accessible and transparent goal. **So we decided that every single decision we made would be guided by a single question: "How does this reduce the transmissibility (r0) of the virus?"**

As we began using this question to guide our decisions, it became a kind of automatic prefiltering mechanism for ideas within and beyond the team. Anybody coming to us with an idea knew the very first question they would be asked, and, in turn, had to give some serious thought to how any new feature or idea would actually help reduce transmission of the virus and save lives.

That singular focus didn't just speed up decision-making, it helped us truly have an impact as we considered aspects that are overlooked. Many team members working on the app initially focused on issues of public trust. The question they asked was, "Will people isolate if the app tells them to," which meant focusing on things like when and how notifications were sent. By encouraging more radical thinking and asking, "What reduces transmission?" you end up thinking about who can GET the app (what platforms, size of app download), and if they can UNDERSTAND what is being said (which languages is it in, is the app copy simple enough to read). We ended up getting mobile phone networks to "zero rate" any app data so it would notify you even if you had run out of data, and created the app in twelve languages. Things that were instrumental in it being a success.

Working at this pace creates stress for everyone, and people are not always at their most considerate. I was getting lots of pressure to do what every stakeholder wanted, however ill thought-out their ideas may have been. I wanted to stop being the "arbiter" or gatekeeper and to make my decisions visible, which is where the single goal became vital. I set up an open prioritization session where any stakeholder, however random, could present their idea by answering three questions: what is your idea, what resources would you need for it, and how will it reduce the transmissibility of the virus?

When these stakeholders came to our open sessions, they had to make their case on the same terms as everybody else. **When you've just heard numerous informed and impassioned cases about ideas that could save lives, it is much harder to argue for something simply because somebody important asked for it.** The singular focus became an easy way to embarrass others into stopping stupid ideas being pitched in the first place. In a few cases, ideas that had first come to me as extremely important were simply withdrawn before being presented in this open forum.

Prioritizing in the open against a single, transparent, impact-level goal allowed us to make the best decisions we could at the speed we needed. It also helped increase inclusivity, and helped us make decisions that would have been overlooked if we didn't focus on the real aim of the app. It's so easy to get distracted by multiple competing goals, which end up diluting the impact you have.

At the end of the day, I think the app was a success. There are certainly things I'd do differently now, but I believe we were able to have a positive impact. According to a paper published in "Nature," the app we built likely averted 1 million cases, 44,000 hospitalizations, and 9,600 deaths in its first year. Building a product is never simple work, but it's important to keep in mind, above all else, why you're seeking to build that product in the first place.

CHAPTER 2

Putting Impact First

A s we discussed in the last chapter, some product teams inadvertently insulate themselves from the broader business by taking on low-impact work that invites neither coordination across teams nor scrutiny from senior stakeholders. But even teams doing work that delivers revenue and growth for the business can be low-impact teams if they don't understand *exactly* what the business needs to be successful on its own terms.

Several years ago, I was brought on to help with a scaling software company that had parlayed a few rounds of early investment into a thriving B2B business. Their product provided digital infrastructure to L&D and content teams, and they had successfully built up a roster of medium- to large-sized customers who used their product to develop internal training, onboarding, and education experiences.

As the business grew, their product teams had no shortage of exciting things to work on. One team was building out an iOS app that would make it easier for their smaller customers to quickly and easily create content. Another team was adding new features and functionalities to the cross-platform "player" that could be used to consume that content. These teams were both in the habit of regularly talking to users, and the feedback they received upon each new product iteration was generally some version of, "Holy wow, this is awesome!"

Meanwhile, the more features these teams shipped, the more the business' most critical metrics seemed to be going up. Revenue? Up and to the right! Number of content creators? Up and to the right! Number of folks consuming that content? Up and to the right!

As I began working with the company, everybody was abuzz with talk of the next big round of fundraising. They had built something *awesome*, and it was only getting more awesome by the day. Surely, the next round of funding would give them the juice they needed to reach the next level of growth and success.

During my first week of working with this company, I set about to get a better picture of what exactly that "next level" looked like. How much revenue were our potential investors looking to see? Did it matter where that revenue was coming from? Were they investing in us as a sustainable B2B business, or a growth business with B2C potential?

Even within the company's relatively small product organization, the answers I received were quite different from person to person. Some folks were convinced that the next round would lay the groundwork for a pivot to a broader B2C consumer base. Others believed that their primary goal was to increase monthly recurring revenue to the point where they would be able to demonstrate a clear path to profitability. **Everybody believed—and could point to metrics proving—that they were contributing to the overall success of the business. But nobody was entirely clear *how much of what kind of success* the business needed, by when, and why.**

During my next catch-up with the company's founder, I shared my conversations with her. I asked if there was a *specific* sense of what the business needed to achieve in order to raise this next round that everybody was so excited about. Was it revenue? Was it user growth? Was it *both*? And in any case, how much? And when?

The founder, who had done a phenomenal job building strong relationships with investors, immediately scheduled a call with me, her, and her last round of investors to get a specific sense of what would be needed for the next round.

During this meeting, we went back and forth trying to get a ballpark figure of what *specifically* would be needed to raise our next round. Through this conversation, two very different scenarios emerged:

If we wanted to raise a BIG round as a growth business with B2C potential, we'd need to see 10,000 monthly active content creators by the end of the year.

If we wanted to raise a smaller round as a sustainable B2B business, we'd need to see $50,000 of monthly recurring revenue by the end of the year.

With this information in hand, we held a meeting with the product org the following week. Their perspective was clear: there was no realistic path to the 10K monthly content creators we would need for that explosive next round. But we *did* have a pretty good chance of hitting our recurring revenue target and forging a path as a sustainable business. By the end of that week, backlogs had been reprioritized. Relationships with enterprise customers had been reenergized. Half-built features had been put on pause. And the company was on track to successfully raise its next round of funding. Which, for the record, it did.

As this story illustrates, a team that makes things go "up and to the right" is not necessarily an impact-first team. In this chapter, we'll discuss how teams and organizations can define impact within the unique context of their own business.

Rethinking "Viability"

This is a book about the steps that individual product teams can take to put business impact at the heart of their work. And yet, this is a *chapter* about business-level goals and milestones that are often well beyond any one product team's sphere of influence. What gives?

To answer this question, let's return to one of the fundamental frameworks of product development. In both product and design circles, individuals and teams are often asked to consider three dimensions when seeking to build successful products: desirability (whether people actually want the product), feasibility (whether the product can actually be built) and viability (whether the product can be built profitably).

Here we run up against one of the most challenging disconnects between product development theory and reality: a product that is desirable, feasible, and generally *viable* might fail completely if it does not contribute meaningfully to the *specific* needs of the specific business building it.

If a company is backed by large amounts of venture capital, then adding more active users might be much more important than driving profitability. If a company is looking to grow sustainably with minimal investment, then finding ways to reduce costs and increase profitability might be much more important than driving growth. And if a company is publicly traded, there might be *specific* targets around growth and/or profitability that have informed shareholder expectations for the year or the quarter.

All of which is to say, **a product team cannot really deliver "viable" work if they don't understand what *viability* means to their particular business**. Understanding what success means to *your* particular business is a necessary first step for any product team that truly wants to prioritize business impact.

? ONE POWERFUL QUESTION

What are the measurable conditions that must be met for our business to be successful at a specific point in the future?

Doing high-impact work starts with defining what exactly the business needs to succeed. But for many companies, success is defined far too vaguely for it to actually influence day-to-day product decisions.

If, for example, a startup has declared its ambitions to "disrupt the industry" or "build something awesome"? That certainly isn't specific enough to drive the kind of decision-making that will allow the business to survive, let alone thrive. Similarly, if a large corporation has provided a list of broadly framed "initiatives" for teams to work on, this in no way guarantees that those initiatives will deliver the results that executives or shareholders are actually hoping to see.

By contrast, if a startup has declared its ambitions to, "Have at least $10K in monthly recurring revenue by the end of July so that we can raise our next round of funding," that makes it clear what product teams should be prioritizing, and what's at stake. Similarly, if a large corporation has promised shareholders that it will expand into three new international markets while increasing its domestic profits by $100M, that gives its employees a sense of what conditions must be met for the business to be successful in the eyes of the broader market.

As these examples illustrate, answers to this question often take the form:

We must have at least [**number**] of [**business metric**] by [**date**] so that [**business-critical milestone**].

Returning to the example at the beginning of this chapter, our initial conversation produced two such statements, depending on which business-critical milestone we were working towards:

> We must have at least **10,000 monthly active content creators** by **the end of the year** so that **we can raise a big round of funding as a growth business**.

> We must have at least **$50K of monthly recurring revenue** by **the end of the year** so that **we can raise a smaller round of funding as a business on its way to profitability**.

Presenting high-level business goals in the form of these simple statements helps us understand both the goals we should be working towards and the stakes of those goals. (Note that for some very large companies, the most relevant goals might belong to a particular brand or line of business within a much broader organization.)

Impact as a Function of Your Business Model

When we talk about "impact," most product people hear "money." And, in many cases, they're not wrong. Money is, by all accounts, certainly an important factor for most, if not all, businesses. (And, conveniently enough, also how most product teams are compensated.) Product teams cost money and, in theory, they should deliver *something* of equal or greater value to the business.

This idea can be jarring for product teams who believe that their primary job is to meet the needs of the *user*, not the *business*. Sure enough, "user centricity" is a powerful rallying cry in product development circles, and for good reason. Simply building what the business tells you to build without considering your users is never a good idea. But seeking to understand the *business model*—how your business sustains itself within the broader market—can help you build things that meet the particular needs of both the business *and* its users.

Several years ago, a very talented engineering leader I knew asked her company's CEO why she didn't hear more product teams discussing the company's *business model*; the fundamentals of how and why the company earns and invests its resources. The CEO was genuinely surprised by this question. For months, he had been fielding complaints that the company was too business-focused and not customer-centric enough. And here was a leader from *engineering* asking for the company to be *more* business focused?

After a bit of back and forth conversation, the engineering leader zeroed in on the source of confusion.

"When product managers say that we are too *business-focused*, I think they mean that we're too focused on the internal workings of our business; the politics, the processes, that kind of stuff. **The business model—how we make money—is** *intrinsically* **customer-centric.** Focusing on the business model forces us to learn more about what our customers find valuable. Otherwise, we can't sell them anything."

This engineering leader had articulated something that many product teams still struggle to believe: **starting with business-level impact in mind doesn't mean you are putting your customers last.** It means that you are putting *the commercial relationship between your business and your customers* front and center, and letting that relationship guide how you learn about and build solutions for your customers.

This conversation also provided a nuanced answer to a question that has hounded the product community since its inception: "Should product managers have P&L responsibility?" (A business-ese way of saying, "should product managers be responsible for the profits and losses their teams generate?") If a company's most important goals involve net revenue, then a

P&L might be the best way for a product team to measure and monitor its contributions, and to open up the possibility of a product team achieving its goals by *reducing* costs like infrastructure and third-party integrations. But if a company's most important goals involve growing its user base, exploring new revenue streams, expanding to new markets, or raising the next round of funding, then managing to a P&L might very well *limit* a team's ability to deliver the desired impact.

Output, Outcomes, and Impact

Putting business-level impact first means fundamentally reimagining the way we think about a product team's responsibilities. Building on a model from Joshua Seiden's excellent book *Outcomes Over Output*, most product teams are responsible for the output they deliver (the things they build) and the *outcomes* that stem from the outputs (the user behaviors that, in turn, create value for the business). Impact—the ultimate value to the business of those changes in customer behavior—is often seen as the purview of executive leadership.

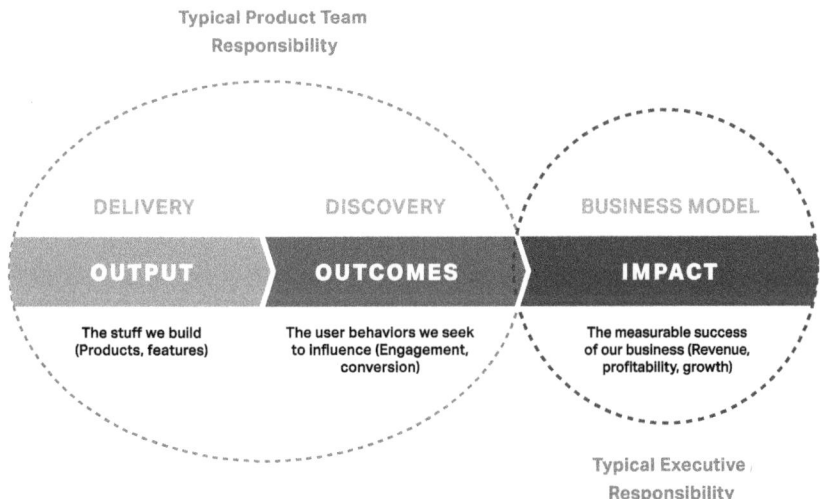

Figure 2-1: Impact, outcomes, and output responsibilities in most product organizations

In theory, this makes perfect sense. Executives *should* be best equipped to align intermediate goals and initiatives with the overall success of the business. But if the last few years of tech layoffs have told us anything, it's that **product teams will ultimately bear the consequences of low-impact work, even if the lack of impact is due to decisions beyond their ostensible area of responsibility.** At the end of the day, a product team's success is inexorably tied to the success of the overall business, no matter how many layers of abstraction are placed between the two. (We'll discuss this more in the next chapter!)

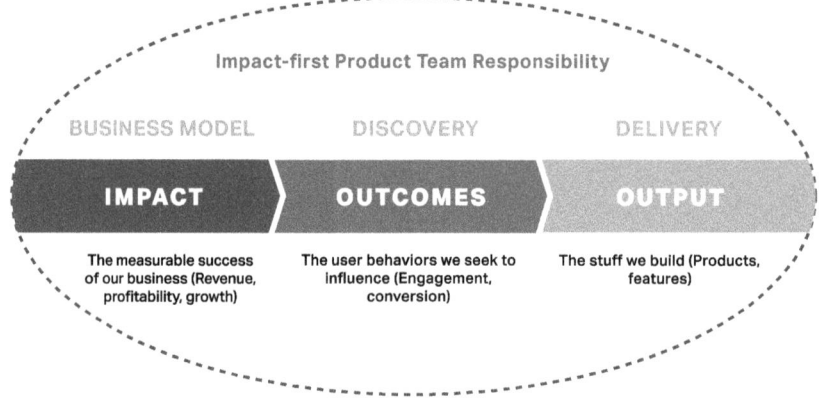

Figure 2-2: Impact, outcomes, and output responsibility for impact-first product teams

If we reimagine this system of output, outcomes, and impact as one that *starts* with impact, then we can better visualize the full real-world purview of a product team's responsibility. This is not to say that product teams should *only* be responsible for impact and should have no outcome-level goals—or, heaven forbid, no actual *output*. **Instead, product teams should be cognizant of the impact, outcomes, *and* output they are seeking to drive—and should always be willing to adjust the latter two in service of the first.**

By way of example, I recently spoke to a B2B startup founder who had found success with a small number of initial customers and was looking to build a scalable, fundable business. He explained to me that he was laser-focused on retaining their existing customers, an outcome-level goal that was guiding the day-to-day output of their small product team. Sure, he acknowledged, simply building what a small handful of customers wanted wasn't the best strategy long-term, but it seemed like the best way to make sure the company would live to fight another day.

That last bit piqued my interest. "When you say 'fight another day,' what do you mean? What would the business actually need to be successful at the highest possible level?"

He answered immediately, "Revenue. Recurring revenue. That's what we need to fund the business, and that's what our investors want to see."

I paused for a moment. "Does that revenue *need* to come from retaining your existing customers? What if you could get a few new customers that stood to bring in *more* revenue?"

"Hmm," he said. "Yeah, I mean, whatever gets us where we need to be. Retention just seemed like it would probably be the easiest way for us to get there."

To be clear, focusing on existing customers *might* wind up being the best way to achieve the company's impact-level goal of growing its recurring revenue. **But giving a product team an outcome-level goal (customer retention)** *without* **a corresponding impact-level goal (recurring revenue) potentially limits that team's ability to pursue** *any and all available levers to ensure the business ultimately succeeds.*

In other words, impact-first product teams don't avoid outcome-level goals altogether; they simply acknowledge that those outcome-level goals can and often *must* shift if there are better opportunities to contribute to the business's existential success. In chapter 6, we'll discuss at more length what these shifts can look like, and how to ensure that they take place when needed.

Making Room for Ethical Concerns Or, Goodhart's Law and Its Limits

Beleaguered product managers love to trot out Goodhart's Law: "When a measure becomes a target, it ceases to be a good measure." In other words, once we set a target, we will do *whatever we can* to juice that target, even if it comes at the expense of our business, our customers, the world at large, or the ostensible reason behind the target itself.

Sure enough, the product world is *full* of examples of teams doing unfathomably shady things to hit their numbers: bank accounts opened without customer consent, emissions results entirely falsified, or the good old-fashioned cooking of the books to create the appearance of success where there is none. I have had many such examples frantically thrown at me when I suggest that being accountable for business-level impact is the only way to align a product team's success with the success of the business overall.

There are, of course, also some powerful counter-examples of technology companies effectively excluding a particular product or initiative from impact-level scrutiny, only to shut down the product, lay off or reshuffle the team, and take a massive write-off years later. (See: Google Plus, Amazon Kindle Fire Phone.)

While these stories might seem different, they all ultimately reach the same conclusion: sooner or later, you can't escape reality. If you're juicing a number to the detriment of your users, your users will figure it out and take their business elsewhere. If you're *ignoring* a number because it suggests that a cool new idea won't work in the real-world market, that real-world market will ultimately have its say.

Allow me, for that reason, to make the following provocative suggestion: **when business-level impact is understood across the organization, it is *easier* to discuss ethical trade-offs, not harder.** Nearly all stories of deceptive patterns, ethical breaches, and ill-advised ideas carried too far all involve teams and individuals working under a cloak of secrecy. They are, in effect, stories of teams taking on clandestine high-impact or low-impact work because the impact of the work being done is *not* something that can be discussed and debated in public. When the trade-offs between ethics and profits are brought into the open, they can be navigated more thoughtfully.

Imagine, for example, that you are working on a team that manages lapsed subscriptions for a streaming video service. You do a little bit of digging, and find that a substantial number of users are paying for subscriptions that they literally *never* use. The ethical course of action would likely be to notify these users that they have an unused subscription, or to suspend payment entirely after a given notice period. But this would obviously come at a cost to the business overall.

Would that cost be worth it? That question is very hard to answer without being able to quantify what exactly "it" is, and whether "it" constitutes a meaningful portion of the company's revenue goals. If, for example, suspending unused subscriptions would only affect .5 percent of the company's yearly revenue targets, it could be a worthwhile gesture of goodwill. Maybe even an opportunity to drum up some good press. If it would affect 20 percent of the company's yearly revenue targets? That's a much more complicated conversation.

These complicated conversations don't always yield wise decisions. Companies can still choose short-term gains over long-term stability, or to chase growth-at-all-costs over sustainable profitability. But these decisions are better made in full view of the organization than under cover of dark and secretive boardroom. And being able to speak the language of business impact means that you can have a say in how your team navigates these trade-offs.

Understanding Company-level Success When It Isn't Clear

In theory, every company should have a clear, specific, and time-bound sense of what success looks like and why. In practice, this is hardly the case. Many small- and medium-sized organizations don't commit to these goals at all. And, as noted earlier in this chapter, many big organizations have so many layers of "goals" articulated in different internal and external decks, documents, statements, and reports that it can be impossible to pin down exactly what success *really* looks like. **If we're supposed to put impact first—and if "impact" is progress towards success—what do we do if nobody has told us what "success" means?**

The short answer is: You're better off having your own point of view on the success of the business than you are waiting around for top-down clarity or throwing up your hands in frustration. The good news is, your point of view doesn't need to be terribly complicated. Depending on the nature of your business, there are a few fairly common impact-level company goals that can be a good place to get started with figuring out what success might mean to the business at large:

Business Type	Common Impact-level company goal	Question to help you get more specific
Early-stage venture-funded software business	Number of daily active users	"What do investors want to see before we raise our next round?"
Growth-stage business	Annual recurring revenue	"What do we need to achieve in order to IPO?"
Publicly traded company	Quarterly earnings	"What commitments have we made to shareholders publicly?"
Nonprofit	Subject area-specific impact goals	"What are the measurable signs that this organization is achieving its stated mission?"

Table 2-1: Common impact-level goals and clarifying questions for different business types

Of course, the above starting points are just that: starting points. Getting a clearer sense of exactly what success means to the business can be quite challenging and nonlinear, especially when different parts of the business might be defining success in different ways. At large companies in particular, this information can feel far-flung and largely irrelevant. But having some sense of the company's specific high-level ambitions can be hugely helpful, even if you are not being asked to consider these ambitions on a daily basis.

These ambitions can sometimes be found in shareholder reports, "town hall" meetings, or even a quick Google search to find press-facing communications. I generally find it helpful to **start by looking for any promises that have been made to critical stakeholders publicly, whether those are investors or shareholders.** (For example, recurring revenue numbers pitched to investors or earnings projections shared in public documents.) We'll discuss how to break down these high-level ambitions in the next chapter but, for now, keep in mind that understanding company-level success can be critically important even if it feels several leagues beyond your immediate sphere of influence.

In Summary

Every team wants to be successful. But an impact-first team starts by understanding what exactly "success" means to their particular business, and why. **Even a basic understanding of a company's funding strategy, business model, and revenue targets can be critical for aligning a product team's success with that of the business overall.** In the chapters that follow, we'll discuss how to break down business-level impact to the team level, and how to estimate the impact of the work our teams consider taking on.

Understanding the Business Model to Improve Product Decisions
Prachi Garg
SVP Product, airport lounge aggregator

When I joined an airport lounge aggregator as SVP of product, we had a lot of different objectives we were tracking progress against. These objectives were at vastly different altitudes; some were revenue related, some were lower-level metrics, some were just features we wanted to build. Making decisions against this wide and varied set of goals was, to say the least, challenging. With unclear and competing goals in place, prioritization was mostly a question of who was advocating the loudest for whatever they wanted to build.

In many ways, the variety of our goals reflected the complexity of our position in the market. Most of our revenue came from B2B partners , who bundled our airport lounge membership with products like premium credit cards. In order for those products to offer value, we had to maintain relationships with independent airport lounges around the world. To further complicate matters, we also offered a membership B2C product direct to consumers. That's a lot of different stakeholders with a lot of different needs, so it makes sense that our company goals would be fragmented as a result.

When I started, my first goal was to figure out how all these different bits and pieces were connected. It turned out that—for what was, in many ways, a complex B2B2B2C setup—our business model was actually pretty straightforward. Every time a member uses one of our airport lounges, it triggers a fee paid by the entity that provided the membership; usually a benefit-provider such as a credit card company. We retain some of that fee, then pass the rest along to the lounge. While partnerships with benefit-providers can also involve revenue shares and commissions, the simplest way to think about it is: the more times our members visited a lounge, the more money we made.

Having our business model laid out in clear and accessible terms had an immediate effect on the way our product teams made decisions. Rather than working against a fragmented set of objectives, we were able to ask a single, simple question: "How do we optimize the number of lounge visits?" (In other words, how do we maximize overall lounge visits while taking into account capacity and experience issues at individual lounges?) There were a number of levers we had at our disposal to answer this question. We were also able to take a fresh look at our app with a much clearer sense of how we would measure its success.

Because we knew how much revenue was associated with each lounge visit, we could also make much better decisions about which opportunities were actually worth pursuing. If something was going to have a negligible impact on top-line revenue, we could deprioritize it. And if an opportunity arose to pursue new membership channels or partnerships, we could measure the size of that opportunity dollars-to-dollars against our efforts to optimize lounge visits from our existing member base.

I've worked with a lot of product teams that are passionate about addressing user problems and driving customer satisfaction metrics. But it's easy to solve and measure these things in little silos and miss the big picture of what value we're providing, and what value we're receiving in turn. **If you don't really understand how the company makes money, you've got a problem**. And when you do understand how the company makes money, you're better positioned to understand and address customer needs holistically in a way that still adds up to a thriving and sustainable business.

CHAPTER 3

Defining Impact for Your Team

E ven when an organization has clear and quantifiable goals, deciding how an individual team can best contribute to those goals is rarely an easy task. Recently, I spent a few months working with one of the core product teams at a financial services company. This company had ambitious, specific revenue goals, and had tasked multiple teams with identifying the areas where they could best contribute to these goals.

The particular team I was working with had been instrumental in growing the company from a single-product offering to a multiproduct platform. But now they were faced with a new kind of challenge. In the past, the business had given them clear instructions for the products and features to build, like a bank account product for small businesses or a spending insights platform for credit card holders. Now they were specifically being told *not* to build anything new. Instead, they were charged with figuring out how they could use the existing platform to help the business achieve its fiscal goals.

At first, the team was outraged by this. "How are we supposed to tell the business what *our team* is going to contribute to something as far-fetched as revenue?! What if our competitors do something that completely blows us out of the water?! Why isn't marketing responsible for this? Or sales?"

The thing is, marketing and sales *were* responsible for this as well. Every team was being asked to think about how they would contribute to the company's overall revenue goals. And—perhaps even more annoyingly—they were being told that "we can't do it alone" was no longer an excuse. They'd have to do it, and they'd have to do it together.

So I sat down with this product team's leadership and began mapping out some of the different ways they could contribute to growing the company's revenue. For starters, there was the obvious option: They could bring in more users. The company already had a widely socialized measure of CLTV—customer lifetime value—or the estimated revenue a new customer might bring to the business over time. With this measure in hand, the team could get a sense of how many new customers they'd need to onboard in order to make a contribution that they felt was meaningful.

One team leader scrunched up her face a bit.

"This just doesn't feel like what this team should be doing. We have a whole lot of marketing and growth expertise in other parts of the organization, and those teams really know how to bring in new users. We've spent the last few years focused on creating this amazing platform. How do we build on that strength?"

I took a moment to zoom out.

"I can't believe I haven't asked you this before," I said, and, truly, I couldn't believe I hadn't asked them this before, "but why is all this platform stuff so important anyhow? How do we measure its success?"

One of the team leaders was ready with a *very good* answer: "All of our studies have shown that multiproduct users are more valuable to the business. In fact, we were able to calculate a higher CLTV for folks who use more than one product." A pause. "Which means that if we could get enough single-product users to become multiproduct users, we could probably make a significant contribution to the company's revenue goals!"

And so it was. We looked at the numbers and figured out how many users we'd need to convert to multiproduct usage to significantly impact the company's revenue goals. We ran it by the product team themselves to make sure it was realistic enough. We ran it by company leadership to make sure it was ambitious enough. And, for the next year, the team had a singular and meaningful goal: turn 10,000 single-product users into multiproduct users.

I'll start chapter 5 with a story of how this specific and impactful goal changed the way the team approached its day-to-day decision-making. But

I'm opening this chapter with *this* story because it perfectly illustrates how **impact-first teams can find a team-level goal that plays on their strengths, informs their decisions, and ties in *directly* to company-level ambitions**. We'll spend the rest of the chapter talking about how to set team-level goals that keep business impact front and center.

 ## ONE POWERFUL QUESTION

What measurable contributions will this team make to be considered a successful part of the business at a specific point in the future?

When I'm brought in to help medium-to-large companies set goals, I usually start by working at the team level. These organizations usually have multiple "teams" working at multiple levels, ranging from three-person "product triads" to thousand-person departments. But, as we will discuss in this chapter, **any team at any level within an organization should be able to understand and articulate how (and how much) its work contributes to the overall success of the business.**

For a team to proactively ask and answer this question themselves puts them in a uniquely strong position, especially when the broader business has specific and critical goals that must be achieved. **Teams that have a point of view on their contribution to the business at large often wind up in the strongest position to understand and navigate the *expectations* of the business at large.** While low-impact teams await marching orders and complain about a lack of clarity coming from senior management, impact-first teams say, "Here are the goals of the business at large, and here's how we plan to contribute to them."

Of course, "plan" is the operative word here. The purpose of answering this question is not to wring certainty from an uncertain world, but rather to understand how an individual team's goals align with the company's overall success. As we'll discuss later in this chapter, this understanding can be particularly valuable for teams whose contribution to the business cannot easily be measured in dollars or users.

High-impact, High-specificity

There is a single fact I've observed over the last decade that has changed my approach to product work more than anything else: **the most effective product teams are those that have team-level goals that are both** *high-impact* **and** *high-specificity.*

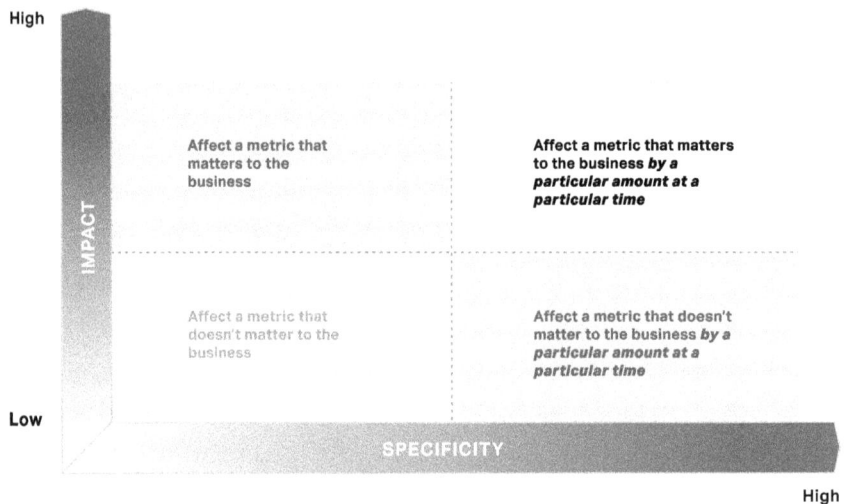

Figure 3-1: Mapping team-level goals to impact and specificity

Returning to the story at the beginning of this chapter, the goal reached by the team in question met both of these criteria. It was high-impact, in that it could be tied *directly* to the company's most critical revenue goals. And it was high-specificity, in that it represented an ambition to achieve a *particular amount of impact* by a *particular moment in time.*

Applying the format we explored in the last chapter to team-level goals, we could articulate this goal as:

> We aim to achieve **10,000** of **single-product users converted to multiproduct users** by **the end of the year** so that **we can contribute 1 to 2 percent to the overall 18 percent revenue growth target that the company shared with its investors.**

After a few years helping teams put these high-impact, high-specificity goals in place, I now feel strongly that **these goals are absolutely critical for teams to make impactful day-to-day decisions.** Regardless of what framework or approach teams use to set their goals, keeping those goals high-impact and high-specificity provides the clarity and guidance that teams need to draw a line from the work they're doing to the success of the business overall. In this chapter, we'll get into more detail about how to set goals that are both high-impact and high-specificity.

Resisting the Urge to Cascade and Control

One of the biggest impediments to impact-first teams is the tendency to break down impact-level goals until they no longer add up to a successful business. The temptation here is pretty straightforward: Big goals can be unmanageable and out of control, whereas smaller and lower-level goals can be more easily and straightforwardly achieved. Revenue goals "cascade" down to usage goals, usage goals "cascade" down to achievable to-do lists, and before you know it you wind up with a failed organization full of teams that are "succeeding" at their individual goals.

How to avoid this dynamic? My favorite approach comes from Christina Wodtke's ever-excellent *Radical Focus*. In this book, Wodtke recommends that each team or individual's goals orbit one level around the company's overall goals, rather than sitting multiple layers down in a questionable "cascade."

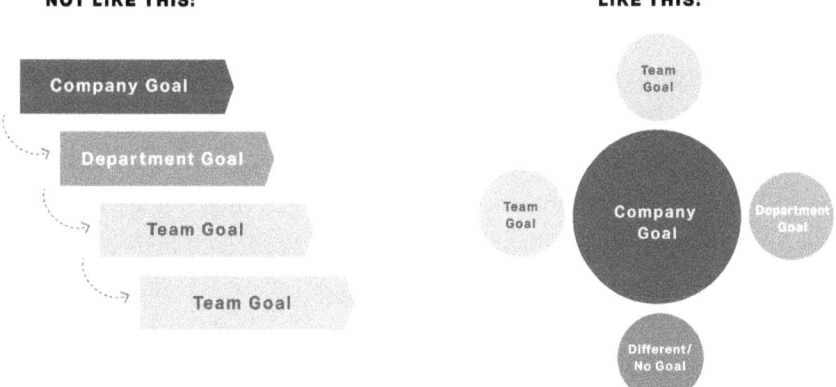

Figure 3-2: Cascading goals vs. orbiting goals, adapted from Christina Wodtke's Radical Focus

Since reading *Radical Focus*, I now ask any team that has cascaded down their goals from other parts of the organization **if and how those goals *add up* to the company level goals**. The answer, in most cases, is, "We have no idea." And that, dear reader, is not a good answer. If each team can ostensibly "succeed" without their efforts adding up to success for the business, then the business is incurring massive and unacceptable risk. These days, I usually ask teams to simply remove any layers of goals that don't clearly add up to success for the business. And what do they wind up with? Well, it looks an awful lot like what Christina Wodtke visualized in *Radical Focus*.

Long story short: **the very things that compel most teams and organizations to continue breaking down and cascading goals should be the signs to *stop*.** I've seen companies waste upwards of six months trying to get a perfect cascade of goals from the tippy top of the organization down to the individual level, only to realize that those individual goals no longer have *anything* to do with the company's success. If we imagine our team goals orbiting around our high-level company goals rather than fitting into a multilevel cascade, we are less likely to waste time and effort creating a false sense of certainty and control.

Staying One Step from Company Goals

It didn't quite hit me until I saw the above visual from *Radical Focus*, but once you see it you can't unsee it: across a number of product success measurement frameworks—from OKRs to "north star" metrics—the most meaningful team-level goals **are no more than one understandable step away from the business' most critical measures of success.**

Keeping team goals for *all kinds of teams* one step away from company goals creates a powerful balance of focus and flexibility. It gives teams the opportunity to strategically narrow their efforts based on their experience and expertise without losing sight of the bigger picture. It makes it easier for teams to adjust course when macrolevel changes take place. And, critically, it means that teams across the organization can align around the same company goals when planning work and allocating resources.

How can you tell if your team goals are one step away from company goals? Broadly speaking, I've found it helpful to approach this question in both quantitative and qualitative terms.

Quantitatively, team goals should be **no more than one mathematical operation away from company goals**. To return to the example at the beginning of this chapter, the team's goal of 10,000 multiproduct users is one operation (multiplying by the increased lifetime value of a multiproduct user) from being expressed in terms of the company's revenue goals.

Qualitatively, team goals should be **no more than one "why" away from company goals.** Again, the team goal at the beginning of this chapter is one compelling, accessible "why" statement away from company goals: "Why are we seeking to convert 10,000 single-product users to multiproduct users? *Because* multiproduct users have a higher CLTV that contributes to the company's revenue goals."

Prachi Garg's sidebar story from before this chapter follows a similar pattern. (You *have* been reading the sidebar stories, haven't you?)

What does it look like when a team's goals are *more than* one step away from company goals? Sometimes, it looks like a handful of disconnected lower-level goals that a team can cherry-pick later on to paint a picture of success regardless of what actually happens. Sometimes, it looks like a set of features to build or tasks to complete without any real *goals* at all. And sometimes, it looks like, "Please refer to slide twenty of our quarterly planning document." When the connections between a team's goals and the company's overall goals are too complicated to be articulated in a single step, they are much less likely to drive impact-first decision-making.

Setting Specific Targets by Evaluating Orders of Magnitude

Now let's talk about specificity. Committing to team-level goals that are one step away from company goals can be difficult enough for product teams. But to truly drive team prioritization and decision-making (which we'll discuss more in chapter 5), teams must commit to *specific* targets and timelines for their goals. For many teams, figuring out what this target should be seems like a near-impossible challenge. But when this challenge is approached with the explicit goal of finding a *good enough* number to drive decision-making, it can be far less daunting.

A few years ago, I was working with the team responsible for building a new version of a long-standing HR management platform. After months of careful consideration, company leadership had decided that the old platform was essentially a lost cause. This team had been given a clear mandate: build a platform that would be worthy of their current users and would help them grow into new markets more rapidly and efficiently.

There was only one problem: they weren't sure *how many users they were supposed to bring onto this new platform, or when they were supposed to do it.* They knew that they were expected to have *something* to show by the end of the quarter, but there was disagreement both within and beyond the team about what exactly success looked like. Was the goal for them to *begin* onboarding users onto the new platform by the end of the quarter? Or to have the entire tens-of-thousands-strong user base migrated? As is often the case in these situations, nobody on the team wanted to volunteer a specific number. They knew *generally* what was expected of them, and it seemed safer to simply see what they could get done and then declare it to be a success.

Unfortunately, this success-by-deferral strategy was making it impossible to actually build anything. Were they building a "minimum viable product" to test and learn from? Or were they building a cutting-edge stunner that would wow users of the old platform? **Without knowing the magnitude of their mandate, they were effectively unable to get anything done.** So I asked them a simple question:

"By the end of the quarter, do we want to onboard 10 users, 100 users, or 1,000 users?"

Silence. Just as I was preparing to ask each team member to write their answer down on a secret ballot (a handy facilitation trick for kickstarting treacherous quantitative conversations!) one of the engineers on the team volunteered:

"You know, I think we could do 1,000. And if we knew that was our goal, we'd be able to say 'no' to a lot of things that don't really get us there."

. . . and there it was. The next day, we got on a call with the company's Chief Product Officer and shared our conversation with her. She promised to give us the support and air cover we needed to work towards our goal. Meanwhile, the team got to work identifying *which* 1,000 users would be first in line to use the new platform. **Having a specific number to work towards unblocked all the decisions that came after; which users to focus on, which features and functionalities they would need, and when those features would be "good enough."**

And here's the thing: 1,000 was by no means a "magical" number. It wasn't the perfect number, or even the "right" number. But it was *different enough from the other numbers we discussed* to compel a concrete and distinct course of action. For that reason, when a team gets stuck on figuring out "how much," I've found it really helpful to think in orders of magnitude: **Is our goal X, 10X, or 100X by a specific date?** What would each approach entail? And what trade-offs would we need to make?

Keep in mind that the answer is not *always* 100X. As we'll discuss in chapter 5, big swings often involve less certainty and can take too long to deliver the needed results for the business. (In fact, the discussion above included about ten seconds of considering *10,000 users onboarded* as a target, only to be swiftly dismissed as both too ambitious *and* too unfocused for us to actually prioritize and deliver effectively.) Evaluating team goals at different orders of magnitude helps us understand what would go into each approach which, in turn, helps us pick a "good enough" number to move forward.

Practical Examples of "Good Enough" High-impact, High-specificity Team Goals

In my experience, the teams best equipped to make impact-first decisions are those who are willing to commit to goals that are "good enough" to influence their day-to-day decisions, rather than engaging in relentless circular debates about the "right way" to set goals at a team and organizational level. After all, the true test of our goals doesn't come until we *use* them to guide our day-to-day work.

What might these goals look like for different kinds of teams? Here are a few examples, drawing on the business models we discussed in the last chapter and applying the high-impact, high-specificity approach we outlined in this chapter:

A product team at a "freemium" B2C startup has as a "north star" goal of achieving a 2 percent increase in daily conversion rate from the product's homepage to free sign-ups by the end of the quarter, as a contribution towards the company's overall user growth goal.

In this example, a product team's goal—conversion rate—is one appreciable step away from the company's overall growth goals. The more prospective users this team converts, the closer the company gets to achieving its top-level goals.

A growth product team at that same "freemium" B2C startup has as a "north star" goal of achieving an increase of 10,000 monthly new prospective users hitting the product's homepage by the end of the quarter, as a contribution towards the company's overall user growth goal.

Taking this example one step further, we could imagine another team that is responsible for bringing new users into the door. We could also imagine that there would be real tension between these two teams' goals; after all, if the growth team succeeds in bringing in TONS of new prospective users, the conversion rate might drop. Being able to navigate this tension one step away from top-line company goals allows both teams to openly explore trade-offs and collaborative approaches. (We'll discuss this more in the next chapter.)

A product team at a growing B2B enterprise has the first key result of its OKRs as achieving an additional $500,000 committed annual recurring revenue *from a particular user or customer type* by the end of the quarter, as a contribution towards the company's overall annual revenue goal.

This approach—in which a team takes responsibility for driving a top-level goal by focusing on a particular customer type or segment—can allow for a good balance of direct impact with more specialized subject-matter knowledge and focus.

A product team at a cancer nonprofit is responsible for driving a specific, measurable change in life expectancy in the following five years.

Remember that "impact" doesn't always mean "revenue." For nonprofits and charities, keeping community-level impact in mind can help drive better prioritization decisions as these organizations pursue more product and technology-based solutions.

As you go about setting goals with your team—whatever framework or approach you use—think about whether you are staying close enough to company-level goals to describe your impact in one step. And remember to keep your goals specific enough to drive decision-making as you set about your day-to-day work.

A Checklist for Setting Team-level Goals

This is where I would love to provide a step-by-step guide to breaking down company-level goals into team-level goals. But the complex nature of real-world organizations means that any such process would likely recreate the very cascade-and-control dynamics we're trying to avoid. (D'oh!) However, inspired (as ever) by the work of John Cutler, I've found it helpful to provide a high-level checklist that you can work through with your team to figure out if you are broadly on the right track to ensuring that your team-level goals put business impact first:

✓ Are our team goals **no more than one understandable step away from company goals?**

✓ Do our team goals **contribute to company-level goals enough to justify the company's investment in the team itself?**

✓ Do our team goals **involve customer and market dynamics and behaviors beyond our immediate control?**

✓ Have we considered **at least two alternate orders of magnitude in arriving at our team goals?**

✓ Have we discussed our specific goals with **the folks in leadership who will be holding us accountable for working towards those goals?**

Of course, this list is by no means exhaustive. But it should give you and your team a good starting point for proactively setting team-level goals. You can find a copy of this checklist online at http://mattlemay.com /impact/resources.

Setting High-impact, High-specificity Goals for Supporting and Operational Teams

There is one category of team that often struggles the most to articulate high-impact, high-specificity goals: teams that do operational and other work that supports internal teams. (Think product ops teams, internal tooling teams, and teams working on things like design systems.) The people who work on these teams haunt me at every talk and workshop I give, popping unexpectedly out of corners with anguished shrieks of, "WE ARE A SUPPORTING TEAM! WE CAN'T MEASURE THE IMPACT OF OUR WORK!"

Sure enough, setting impact-level goals for supporting and operational teams can be particularly challenging. Unfortunately, most teams are in the habit of setting their own goals in isolation from other teams, which makes *absolutely no sense* for teams whose impact will ultimately be realized by other teams' work. **For supporting and operational teams to effectively set impact-level goals, these teams usually need to be in close conversation with the teams they are supporting.**

These conversations can help supporting teams transform their transactional relationships with user-facing product teams into deeper partnerships. Many supporting teams I work with complain of feeling like "ticket-takers" fielding constant volleys of urgent and ill-considered requests from teams that don't really understand their capacity or their capabilities. But early impact-level conversations can preempt this dynamic by prompting teams to better align on how they can help each other achieve high-level goals before there are specific bits of work to request.

Here are a few questions I've found helpful for supporting teams as they initiate these conversations with user-facing product teams:

What is the overall impact your team is seeking to drive for the business?

How could our team's support change the amount of impact you have?

How could our team's support shift the timelines in which you deliver that impact?

The conversations that follow from these questions can accomplish two critical things for both supporting teams and their partners across the organization. First, they can help supporting teams articulate their business-level impact as a function of the additional impact they can drive through partner teams. Second, they can help move the relationship between these teams from ticket-taking to strategic partnership. By having these conversations, supporting teams are able to prioritize their time and effort based on what will be most impactful, not who is yelling at them the loudest at any given moment.

When Team and Company Goals *Aren't* Aligned (Don't Panic!)

As we discussed in the last chapter, "success" for a business may mean more than just making money. Early-stage startups might have specific goals around growing their user base that they must achieve to raise the next round of funding. Large companies might have longer-term ambitions that require thinking outside of the company's current business model. Some common difficult-to-align team goals include:

- Delivering against a specific promise (such as expansion into a new market) that has been made to investors or shareholders
- Exploring potential new revenue streams and business models to ensure long-term survival and success
- Creating positive buzz and press for the company by doing high-profile work that will draw industry attention

I've seen many teams tasked with responsibilities like these in a perpetual state of slow-motion panic because their goals don't clearly align with any obvious revenue or growth targets, only to learn later that the company at large had *zero* expectation of their work actually aligning to such targets.

And while the end result may have been relief for the team, the preceding confusion was massively counterproductive. It made it harder for the team to focus on work that would actually help them be successful in the eyes of the business, to say nothing of the stress it caused the actual humans on the team.

If you suspect that your team's contribution to the business does not line up to business-level targets, make sure you've had that conversation openly and explicitly with team and/or company leadership. "We are focused on something else" can be an acceptable answer if and only if it is an answer you know and understand—and has reasons behind it that you know, understand, and can articulate.

In Summary

The temptation to break down high-level, market-based goals into easily controllable nuggets runs deep among teams, individuals, and organizations. But the second we give ourselves the illusion of control, we lose touch with the reality of our business—a business that depends on customer behavior and market dynamics that usually shift and evolve faster than we can keep up with them.

Delineating our individual team's contributions to the overall success of the business is by no means an easy task, nor is it an exact one. But by staying one step away from company goals, exploring different orders of magnitude, and acknowledging that we need only a "good enough" number, we can unblock our team to make decisions that put the success of the business first.

CHAPTER 4

Impact in the Middle

▬▬▬➤

As we discussed at the beginning of this book, most of the writing about product development focuses on "the middle." From strategy to discovery to problem scoping, there is a *lot* of work that goes into the middle, and many phenomenal resources out there to help teams and individuals better navigate it. But regardless of how your particular team and organization approaches the steps between high-level success and day-to-day decision-making, keeping business impact front and center can make these steps more coherent, effective, and efficient.

Several years ago, a client of mine was at an all-too-familiar impasse. The company had invested in a major overhaul of its research and discovery practices, hiring skilled user researchers and putting its product teams through well-received training in the art of working together to discover and understand customer needs. And yet, once these trainings had concluded, things had more or less gone back to the way they were before: product teams pitching new features to executives and then pushing those features out the door with minimal research or discovery efforts. The newly hired researchers were frustrated and demoralized, trying to share user insights with product teams that were "too busy" to take them seriously. And the product teams themselves were left to conclude that, while the company may have talked a big game about being "customer centric," they were (and always would be) just another feature factory churning out executive-mandated widgets.

As grumbling and disappointment spread through both formal and informal channels, the company soon found itself embroiled in one of the most common debates among modern product teams: **how much discovery, exactly, should we be doing?** While this debate didn't help the company achieve any of its goals, it *did* give literally everybody at the company something to be unhappy about. Researchers and research-minded product managers could complain about how the organization wasn't doing nearly enough research or discovery. Folks whose success hinged on actually getting new features out the door could complain about how certain people in the organization were pushing for way too much research and discovery, even when the right thing to build was painfully obvious. Nobody wins!

For my part, I seemed to have no trouble finding new and novel ways to make things worse. Every attempt I made to strike a "happy medium" seemed to make everybody appreciably *less* happy. My conversations with researchers quickly devolved into ineffectual commiseration, and my conversations with product teams seemed like a waste of everybody's time.

And then, just when I thought things couldn't get any worse, the company's CEO swooped in with her own morale obliterator: She had been playing around with the product's new user experience and found it to be confusing and overcomplicated. Based on conversations among the leadership team, she now believed that streamlining the new user experience could increase sign-up rates by at least 5 percent. And given the company's overall timeline and revenue goals, she expected to see that increase in three months' time.

The immediate reaction among product teams was one of outrage and indignation. This reeked of "command and control" behavior! Why was the CEO messing around in the weeds of the new user onboarding process to begin with? And how *dare* she set such an arbitrary-seeming goal without consulting the teams who would be responsible for delivering against that goal?

Despite several outraged complaints to middle management, the message was clear: The product org at large was responsible for delivering a 5 percent increase in sign-ups by the end of the quarter. Teams would need to reevaluate their priorities and shift their resources. They would need to get creative and make substantial changes to critical parts of the application. And at the end of the day, they would all be accountable for an outcome that was solidly outside the collective control of the entire product organization, let alone a single product team.

Once the outrage began to subside, something interesting happened. As the reality sunk in that product teams would be evaluated against outcomes beyond their control, the way those teams engaged with researchers began to change. **The question of "how much discovery should we do?" now felt abstract and irrelevant.** In its place were much more immediate and urgent questions, such as, "Why are folks getting stuck in the sign-up flow," "Is there a particular type of user we're losing more than others?" and even "Are there any parts of the onboarding experience that are unnecessary?"

To my genuine surprise, the very same "manufactured urgency" that had initially outraged product teams started forcing them to get closer to their users, to better understand the company's overall business model, and even to work more closely and collaboratively with other functions and departments. The impasse had been broken, not by agreeing on the "right" amount of discovery, but rather by committing to a specific impact-level goal that made user research an essential factor for success.

Impact First at Every Step

Regardless of its particular shape, "the middle" serves two related purposes in most product organizations. First, it connects the ambitions of the business with the goals and needs of users, giving life to the value exchange at the heart of any company's business model. Second, it begins to put some shape around *how* a business plans to achieve its ambitions, by defining particular strategies, initiatives, and/or "bets," and then breaking those down into more clearly defined problems and opportunities that a product team can address in its day-to-day work.

When approached efficiently and effectively, these steps can bring an organization closer to its customers, codify a product's unique value to those customers (shout out April Dunford's excellent *Obviously Awesome*), and streamline day-to-day decision-making. But, as with the multilayer goal cascades we discussed in the previous chapter, **"the middle" also runs the risk of creating distance and disconnection between a team's day-to-day work and the overall success of the company.**

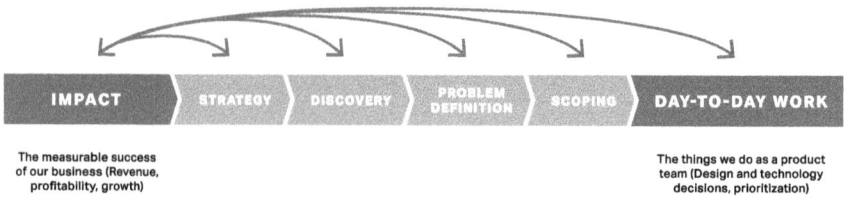

Figure 4-1: Impact at every step through a generalized "middle"

For that reason, I advise the teams and organizations I work with to **keep business impact front and center at *every* step of "the middle," whatever those steps might be.**

In this chapter, we'll take a look at what it means to keep impact first as we walk through two critical aspects of "the middle:" strategy and discovery. Again, the ideas and suggestions captured in this chapter are designed to be helpful for all product teams and organizations, whether they are working in "bets" or "initiatives," whether they are conducting continuous discovery or interfacing with a dedicated research team, and whether they are working with a formalized "strategy" or figuring it out as they go along.

Strategy and Impact

Strategy is, understandably, one of the most hotly debated subjects in the product management world. A quick Google of "product strategy" yields numerous definitive-sounding and largely contradictory definitions, templates, and convincing case studies.

In a 2022 interview with McKinsey (and I promise this is the only time McKinsey will be mentioned in this book!) author of *Good Strategy/Bad Strategy* Richard Rumelt defines strategy as, " . . . how you overcome the obstacles that stand between where you are and what you want to achieve." I quite like this definition, as it speaks directly to impact ("what you want to achieve") and also leaves room for businesses to address the unique challenges they may face based on their business model and position in the market.

When it comes to company-wide strategy, some organizations take a "bottom-up" approach where individual teams can propose strategic initiatives and efforts. Many *many* more take a top-down approach, in which company leaders or a centralized strategy office publish a much-ballyhooed strategy on a fixed (usually yearly) cadence. In my experience, the "strategies" produced in these contexts are often host to a mix of impact, outcome, and output-level goals. Some are concise and actionable, more are overstuffed and feel like the product of PowerPoint design by misaligned executive committee. The more complicated these strategies are, the more likely they are to replicate some of the "cascade and control" dynamics we discussed in the previous chapter. And the more vague and contradictory they are, the more likely they are to be weaponized by folks who dismiss anything they don't like as "off-strategy."

It's no surprise, then, that individual product teams tend to have *feelings* about any strategy they are handed, even if they ostensibly played a role in shaping that strategy. In most cases, product teams are not well-positioned to litigate why and how a company strategy might be muddled or ill-advised. **But product teams with impact-level goals *are* consistently well positioned to make a case for the work they're doing, even if it doesn't align perfectly with somebody else's interpretation of company-wide strategy.**

Returning to Rumelt's definition, the whole point of a strategy is to help us figure out how we are going to achieve the company's most important goals. When a product team can articulate the value of their work in terms of those goals, they are much less likely to get bogged down in theoretical debates about what makes a strategy "good" or "bad." And because impact-level goals usually involve an exchange of value with customers in the broader market, debates at this altitude are more likely to be settled through research rather than conjecture. Which brings us to . . . .

Impact-level Goals as Discovery Catalyst

The process of *discovery*—learning about our users to make sure that our products actually address their needs and goals—is, thankfully, incredibly well-documented. This book is not going to retread that well-trodden territory. If you haven't already, I'd strongly recommend reading Teresa Torres's *Continuous Discovery Habits*, the aforementioned *Outcomes over Output* by Joshua Seiden, and Melissa Perri's *Escaping the Build Trap*.

When we have high-impact, high-specificity goals like those we discussed in the previous chapter, we are able to approach broader discovery efforts with better focus and clarity. Because we have specific timelines and targets, we can prioritize customer segments and discovery methods that will give us the information we need in a timely manner. And because we are keeping high-level business impact front and center, we give ourselves the flexibility to uncover previously unseen solutions rather than validating our existing plans.

Best of all, because we have given ourselves time and target-based accountability for customer and market-based goals, we've created some real urgency around learning things from the broader world. The **discomfort we feel when we hold ourselves accountable for specific goals outside of our immediate control can be the most powerful motivator for teams to actually *do* meaningful discovery work.** Having impact-level goals in place turns discovery from an abstract concept into an essential part of delivering success for the team and the business.

? ONE POWERFUL QUESTION

Who needs to do what for us to achieve our impact-level goals—and *why* would they do it?

As we discussed in the previous chapter, high-impact goals are outside of a team's immediate control. In order for us to achieve these goals, *somebody* other than ourselves—our users, our customers, and/or a particular subset or segment thereof—needs to do *something*. That something might be signing up for our service. It might be purchasing something from us, or purchasing it at a different price. Whatever it is, it isn't something we can simply choose to *build*.

It is this very lack of control that should compel "the middle" to do what it is intended to do: unite business goals and customer needs in a sustainable and actionable plan. Creating such a plan often starts by looking at the side of the business model where the most uncertainty lies: **who (other than ourselves!) has to do *what* for us to achieve our impact-level goals? And, perhaps most important, *why* would they be inclined to do this thing?**

By way of example, I once conducted a research project for a technology events company determined to increase ticket sales for its conference workshops. This company had invested heavily in making sure that these workshops addressed the latest technology trends, but was still struggling with overall sales numbers.

My colleagues and I helped them set a specific goal of an overall 30 percent increase in workshop sign-ups for their next conference. This specific goal helped us focus on two vital discovery questions:

Who signs up for technology conference workshops?

How do they choose which workshops are worth signing up for?

Most of the company's prior efforts had been centered around a single core assumption related to the second question: that it was the *content of these workshops* that would determine their overall success. But having a specific and time-bound impact-level goal—a 30 percent increase in sign-ups *by the next conference*—forced us to zoom out from that assumption. Most of the conference's program had already been decided. If we wanted to achieve our specific impact-level ambitions, we no longer had the luxury of hiding safely behind our assumptions. **We *had* to learn about the goals, motivations, and behaviors of our potential customers or we were destined to fail.**

What we learned was quite surprising and extraordinarily valuable. For starters, we got a much more nuanced understanding of *who* approved the budget for conference trainings and workshops and *how* these approvals played out at many medium-to-large organizations. Beyond that, we got an actionable sense of the practical concerns and actual *search terms* that folks were taking into account when deciding how to utilize their budgets. **Armed with this knowledge, we were able to make a series of relatively minor changes to wording and messaging that helped us achieve a goal that had initially seemed grand and daunting.**

Keeping All Levers in Play

As the above story illustrates, the most impactful things to do aren't always the most elaborate or exciting. Often, they aren't even things that teams "build" at all. As teams go about the work of learning about the goals, needs, and problems of their users—and exploring potential ways to address those goals, needs, and problems—keeping business impact in mind can help teams remain open to potentially unexpected and unexciting things that could contribute meaningfully to the company's overall success.

Indeed, **keeping impact in mind throughout "the middle" makes it easier for product teams to keep all potential levers and solutions in play, regardless of the specific processes they're using or steps they're taking.** Here are a few types of opportunities that the impact-first teams I've worked with have identified in "the middle" that might have otherwise gone unexplored:

Entirely new customer segments and revenue streams

Keeping impact goals front and center can help teams identify whole new groups of current and prospective users to reach, and new ways of exchanging value with them. For example, I once worked on a discovery project where, a few weeks into interviewing users, it became clear that our biggest opportunity to reach our specific user growth goals existed outside of the "power user" segment we were initially focused on. Our high-impact, high-specificity goals provided us with the urgency (and the air cover) to change our approach and identify new growth opportunities that we would have otherwise missed.

Operational and back-end changes

When I worked for a very cool music startup, the most impactful thing my team delivered involved connecting our display advertisements to a new back-end system. This was, to say the least, not the most interesting or exciting thing we could have worked on. But as the team explored multiple ways to increase our revenue per user, it eventually became clear that this work—which represented no meaningful change in the actual user experience—was actually the most important thing for us to deliver because it was most likely to help us achieve our high-impact, high-specificity revenue goals.

Minor changes to copy and messaging

As the story earlier in this chapter illustrates, small changes to the way we communicate with our customers can deliver big results. Keeping impact goals front and center can help us identify and prioritize these opportunities, even when they tend to fall outside of the way we scope and build "features."

Subtractions and deletions

Product teams are rarely incentivized to *subtract* elements of a product. New features are exciting, massive "redesigns" are celebrated, but subtractive changes evoke more scrutiny than praise. Which is a shame since streamlining the user experience can make that experience *way* better and, in turn, deliver a much greater impact. When teams can point directly to that impact, they are much more likely to generate momentum around subtractive approaches to product development. This is particularly true for teams working towards profitability goals that can be achieved in part through reducing costs or increasing efficiency.

For an example of how high-impact, high-specificity goals can drive subtractive product decisions, let's revisit the company we met at the beginning of this chapter. After a few weeks of hand-wringing and uncertainty it became clear that, yes, product teams were *actually expected* to drive a significant increase in new user sign-ups. And as that reality sunk in, those teams began to look beyond their usual habits of pitching, scoping, and building new features.

Through this expanded aperture, one team noticed that there was one *particular* step of the sign-up process that didn't make a lot of sense. Rather than simply signing up for a single plan, new users were asked to put one or more plans in a "cart" and then go through a "check-out" process. This team pulled some data and learned that a *tiny* percentage of prospective users were adding multiple plans before checking out, while many *more* users were abandoning their single-plan carts before completing the check-out process.

In the absence of high-impact, high-specificity goals, there is rarely much incentive to rework a part of the product that touches on multiple teams' areas of responsibility and has been working *just fine* for years. But this team saw an opportunity that could get them *all the way* to the company's most important goal, and they took it. Sure enough, getting it out the door required coordinating across a number of teams and fielding some challenging questions from leadership. But when all was said and done . . . it worked. By removing one extraneous step, one product team was able to increase new user sign-ups by more than 3 percent in one month, getting the company more than halfway to a once insurmountable-seeming goal.

Shifting Forward Cross-Team Conversations

One reason that many organizations still favor a cascade-and-control approach to goal-setting, strategy, and execution is that it promises a world where each team's work can be perfectly carved out without any pesky cross-team "dependencies." Indeed, much of the last twenty years of software development best practices have been about breaking down big things into smaller things that can ostensibly be tackled by nonoverlapping "autonomous" teams.

One of the main criticisms leveled against high-impact goals is that they often undermine this ideal by holding individual teams accountable for results that they can *only* deliver by working closely with *other* teams. But this criticism begins to fall apart when we recognize that the best products feel like seamless and cohesive experiences, not disconnected grab-bags of perfectly nonoverlapping features and "product areas." **Oftentimes, the goal of eliminating tension between teams is fundamentally at odds with the much more important goal of minimizing the friction experienced by users (and, in turn, maximizing the business impact of that experience).**

When teams put impact first at every step of "the middle," they are more likely to manage this tension at *every* step of "the middle," rather than waiting until they have committed to work riddled with unseen tactical dependencies. And along the way, they are much more likely to identify ways that they can work *together* towards achieving high-level goals for the business, rather than competing with each other for resources and recognition.

Of course, this doesn't mean that such coordination will be easy or straightforward. As more teams begin prioritizing against impact-level goals, it is not uncommon for multiple teams to want to make slightly different changes to the most well-trafficked and important parts of a product such as its onboarding flow and home page.

Imagine, for example, the all-too-common scenario of two feature teams fighting over placement on a popular app's homepage.

In our first scenario, each team has the **lower-level goal of** feature usage. The success of each team is measured by how many people use the new features they are respectively building. In this case, the two teams find themselves chasing the same prize: the most prominent available placement on the home page, in the hopes of driving the most clicks and engagement. Are all of those clicks equally valuable to the business? Neither team has much reason to care. They will be evaluated on the ability to push their own features, and push they will, even if it comes at the detriment of the overall user experience.

In our second scenario, these teams have **impact-level goals**, one step away from the company's overall revenue goals. One team is responsible for maximizing the lifetime value of customers by building new features that require additional payment. The other is responsible for minimizing costly churn by building features for basic subscribers that keep up with the competitive landscape. Whose new feature should receive more prominent placement?

There is no obvious answer here, but there are questions that can help us resolve this issue to the overall benefit of the business, such as:

- How many of each type of user is likely to encounter this feature on the homepage?
- What is the likelihood of each feature delivering its respective impact?
- Are there other ways of achieving the desired impact that might fall across or between both teams' responsibilities?

These are tricky questions, but because they can be answered on the same terms across different teams, they are more likely to be resolved in a way that benefits the business overall. Furthermore, they can be resolved in a way that is not entirely zero-sum. Perhaps the first team's feature is best surfaced to users who have already paid for one or more extra features. Or, perhaps the second team's feature is best merged with an existing feature or functionality of the application. The sooner teams have these conversations—and the more approaches they are able to keep in play—the better the outcome is likely to be.

Every organization has its politics, turf wars, and battles over whose team "owns" what. **But by keeping team-level goals as close as possible to shared company goals, these conflicts can be resolved much more effectively.** Teams with impact-level goals are much better positioned to compare their different approaches in terms of their overall contribution to the company's success. Teams with lower-level goals are much more likely to find themselves in zero-sum conflicts that fail to deliver success for the business and fuel the Low-Impact Death Spiral.

In Summary

This is a short chapter for good reason: "the middle" is the most well-documented part of the product development process, and one of the most hotly debated as well. But **when teams are accountable for delivering against goals that *require* up-to-date knowledge of customers and markets, then "the middle" can do what it does best: connect the business with its customers, and provide some structure for teams to figure out how they will achieve their goals together.**

Introducing Impact Metrics at a Legacy B2B Utility Company

Debbie Kite and Nicole Gray

Product and UX consultants, public utility company

We came in as consultants, so we were used to receiving a list of features to build. But this time, we had a partner within the company who recognized the value of setting clear goals and expectations. Working with him, we set about to unpack the "why" behind the features we were asked to build, and express it in clear business terms.

It turned out that the features we were given were broadly intended to reduce the flow of general help requests coming in through customer support channels. These requests were overwhelming customer support agents, to the point that more than 7,000 of them a year were simply rejected, requiring them to be reformatted and resubmitted by the end user. Company leadership figured that we could address this problem by, among other things, building out a robust new "learning center" to help customers answer their own questions.

Given the terms of our engagement, we were not in a position to say, "No, we won't build that." But we _were_ able to work with our internal partner to draw a solid line between the work we were asked to do and the potential impact of that work. This was made possible by the fact that our internal partner—who wasn't technically a product manager but certainly knew how to act like one—had started the project with a "benefits case" laying out what success for this project would look like for the business at large beyond simply building the proposed features. Specifically, he reasoned that if we could reduce those 7,000 rejections by 95 percent over 5 years we could make a quantifiable impact, reducing the company's operating costs by a specific projected amount.

With this goal in place, we were able to open up a bit of space between the work we were asked to do and how the company measured success. Within that space, we were able to conduct some high-quality user research to better understand why and how customers were sending those general help requests, and to test some interactive prototypes of the features we had been asked to build. Unsurprisingly, we found that adding more features didn't necessarily make it easier for folks to find what they were looking for. In fact, it sometimes made it *more difficult* for folks to complete the tasks we were walking them through.

We couldn't exactly take this research back to the company and say, "We're not going to build any of the things you told us to." But we could reshape the features a bit until they were more likely to deliver against the goals we could set. And because we had set those goals in the first place, we could make a case for these new approaches in a way that clearly benefited the business overall. The "learning center," for example, became a simple FAQ page that helped customers find their way more quickly and efficiently. Which, in turn, meant that the business was more likely to achieve its goals and reduce its costs.

While we'd love to say that this experience magically transformed a legacy utility company into a customer-centric digital product machine . . . that just isn't the way things go in the real world. Different businesses have different concerns, different pressures, different relationships with their customers and their shareholders. For many utility companies, their customers don't have a ton of choice, so operational efficiencies are of the utmost importance. By framing up the goals of our work in terms of operational savings, we were able to open up space for the company to better serve its customers. Since then, they've slowly but surely started to measure and monitor the success of their work beyond the simple fact of having launched new features. **Meeting the business where it is and understanding their needs and concerns greatly increases the likelihood of you making real and lasting change.**

CHAPTER 5

Estimating Impact

||||||||||➡

Having clear impact-level goals can help guide discovery and delivery ideas from ideation through execution. But the closer we get to actually *delivering* something, the farther away those impact-level goals become and the easier it is for us to lose sight of them. Let's pick up with the product team we met at the beginning of chapter 3. After aligning the team's strengths to the company's ambitions, they chose a single, team-level goal: turning 10,000 single-product users into multiproduct users by the end of the year. This goal would deliver a meaningful contribution to the company's overall revenue targets by way of increased customer lifetime value.

A few weeks after agreeing to this goal, one of the product managers on this team reached out to me for help. His team had been considering three different bits of work, *all* of which seemed like they would be valuable to both the business and its users. Even with a solid team-level goal in place, he feared that they were slipping back into old habits of prioritizing work by opinion and conjecture.

"OK," I said, "so there are three bits of work you're considering. **How many multiproduct users do you think each of those could give us?**"

A familiar fear swept across the product manager's face. "I mean, I don't really know, *exactly*. We've been superfocused on validating the ideas themselves, and I think they would all be valuable for converting single-product users to multiproduct."

I conjured the most soothing voice I could. "Don't worry about it. You don't need to know exactly! Let's just walk through each of them. Which users are we seeking to convert? Where are they in the product? And how many of them are there?"

The first bit of work, it turned out, would aim to convert users who were already "superusers" of a single product. These users, the product manager figured, would be the most receptive to an upsell message.

"That makes sense to me," I said. "But how many of these single-product 'superusers' are there?"

Again, that look of creeping fear. "I don't know. **I don't have access to the data to give you an exact answer.**"

"Then give me an *inexact* answer. Humor me. Roughly how many are we talking about here?"

"I don't know. Probably . . . about a thousand?"

"OK, so that's about a tenth of our overall goal. That sounds pretty worthwhile to me."

On to the second bit of work. This one involved making it easier for users to convert from a particular landing page in the web app. This, the product manager explained to me, was a no-brainer: The page was well within his team's purview, and there were some obvious and easy ways to improve it.

This time, he anticipated my follow-up question. "That said," he offered, "it's probably not as big an opportunity. I'm guessing that no more than a hundred users will hit that specific web landing page this quarter."

"If that's the *most* we could do," I offered, "then we probably shouldn't spend much more time thinking about it."

And finally, the third bit of work. This one involved making substantial changes to the onboarding experience for *all* users. It would require a lot of coordination, a lot of time, and potentially a lot of messy negotiation with other teams who had their *own* high-level goals to achieve. But the size of the opportunity was potentially huge; The team's entire multiuser goal could be achieved or exceeded by making these changes to the highest-stakes and highest-traffic part of the application.

"Honestly, I'm not sure if we'd be able to get everybody on board to make those kinds of changes. There are lots of teams looking at the core onboarding experience as a way to achieve their goals."

"That's OK," I said. "Let's take out a piece of paper and capture where we are. It might not feel that way, but I think we've made a LOT of progress."

	Max Users	What next?
A	1,000	Continue discovery
B	100	STOP
C	10,000+	Talk to other teams — what are their goals?

Figure 5-1: Capturing potential impact and next steps for three different bits of work

The resulting document captured the estimated maximum impact of our three options, and what to do next with each one. Option A seemed well worth continuing to pursue because it could deliver a meaningful number of multiproduct users, and was relatively likely to succeed. Option C required coordination with other teams to see how their goals and our goals could be combined to best contribute to the company's overall revenue goal. And option B? Well, *let us never speak of it again.* There were still a lot of important decisions to be made, and a lot of follow-up conversations to be had around Options A and C. But on this one sheet of paper, we had drawn a straight line between two very important things: **what the impact of each piece of work could be, and what we were actually going to *do* about it.**

And perhaps more important, we had taken one third of the team's current decision-making overhead off of their plate entirely.

? ONE POWERFUL QUESTION
What is the *most impact* we could reasonably expect from a given piece of work?

The above story doesn't have a neat and tidy ending. In the real world, very few prioritization discussions do. If they did, product work would be easy and product teams would be chill and relaxed. (Spoiler alert: they are not.)

Even in the absence of a neat and tidy ending, realigning our prioritization around impact had two massively valuable outcomes. First, it clarified the trade-offs between options A and C: one more likely to succeed, the other more ambitious in its potential impact. Second, it gave the team permission to **stop wasting its time considering work that had no chance of delivering meaningful results in the first place.**

Which brings us to this chapter's powerful question. Just as different teams take different approaches to working through "the middle," different teams take different approaches to prioritizing and evaluating the work they plan to build. But a truly impact-first team starts by asking for any solution they are exploring: **If this thing were reasonably successful, would it be worth doing?** In other words, if we were to spend weeks validating this solution, further weeks building it, and yet more weeks releasing and iterating upon it, would that investment of time be worth the *most impact* we could reasonably expect?

Starting from a well-reasoned best case scenario allows us to quickly evaluate whether an opportunity is worth pursuing, before we do in-depth validation or documentation. Here, the importance of having specific and measurable definitions of team-level success becomes manifestly clear. If we don't know what, specifically, success looks like, it is *very* hard to know whether any particular piece of work is going to contribute *enough* towards that success to be worth the time and effort to explore, scope, and deliver it.

Derisking against Impact

As we discussed in the last chapter, the vast majority of product management literature is about the process of going from outcomes to output, from opportunities to solutions, from customer problems to working software: "the middle." There are numerous phenomenal books, articles, and podcasts where you can learn more about discovering, testing, and validating the things your team could actually *build* and *deliver* to achieve those goals.

Nearly every team goes through a slightly different set of steps in this well-documented process. Some identify "opportunities" and then discover "solutions." Some break their users down into different groups by "persona" while others identify persona-crossing "jobs to be done." Long story short, most teams go through lots of steps to get from, "What does success look like?" to, "What should I be working on after I finish this next cup of coffee?"

I don't have a strong opinion about which of these approaches is generally best. (Though at the minute, I'm finding Martin Eriksson's Decision Stack particularly useful.) I've worked with organizations that have used any and all of these approaches effectively, and organizations that have ineffectually gone through the motions of those very same approaches. Because regardless of the steps an organization takes to get from impact-level goals to day-to-day work, they incur the same compounding risk: **The farther from impact they get, the more likely they are to follow an intermediate goal or plan that might be outdated, off-base, or otherwise fail to deliver success for the business.**

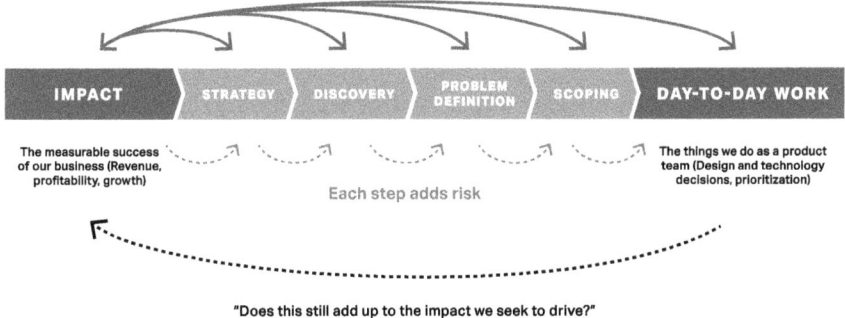

Figure 5-2: Checking in against high-level impact after incurring risk at every step of "the middle"

The team whose story we revisited at the beginning of this chapter went through their own well-trodden process to get from goals to user problems to opportunities to solutions. But, as that story illustrates, they had drifted away from their impact-level goals by the time they began prioritizing specific solutions. **If each of these steps were perfectly aligned, then understanding the impact of our work would be *really easy*.** You'd just work your way back along each step you took until the beautifully manicured breadcrumb trail of product excellence spelled out *exactly* what each bit of work would deliver for the business.

In practice, though, it's never quite so easy. Each step taken from goals toward day-to-day work adds ever-compounding risk. Risk that you may have overlooked a critical uncertainty or assumption, risk that your target customers' needs and goals had already changed—and, most pressing of all, risk that your carefully crafted strategic initiatives and priorities are too complicated and convoluted to drive impactful decisions.

Losing track of impact as we begin prioritizing and building is a common problem for product teams. Drawing a straight line from high-level impact to day-to-day work is by no means an easy task. But for impact-first product teams, it is a necessary step to ensure that success is not just defined, but delivered.

Getting Started with Impact Estimation

The activities described in this chapter are often broadly referred to as "impact estimation" in product development lingo. It is, somewhat confoundingly, something that product teams are often asked to do *after* assembling a complete roadmap of the work they plan to deliver. Which, thankfully, is not a problem for teams who have kept impact front and center every step of the way.

There is no single way to approach impact estimation that works for every team. The nature of work being done, the team's particular goals, and the company's business model can all affect how we approach this task. **But it is important to remember that the goal of this task is to *estimate* impact, not to *perfectly and precisely predict impact*.** As we've discussed, any work that is worth doing involves some degree of uncertainty. Impact estimation gives us a way to acknowledge that uncertainty while doing enough "back of napkin math" (like the math we did at the beginning of this chapter) to anchor our conversations and decisions.

For teams that are not accustomed to estimating impact, getting started can be quite challenging. Especially if they have only undertaken this activity as a reactive response to unwanted executive scrutiny. I've found that the below questions can provide a good starting point for a broadly applicable approach to estimating impact:

How many people is this work going to reach?

As the story at the beginning of this chapter illustrates, many things that seem like "quick wins" are quick precisely because they won't reach enough people to have a meaningful impact. Asking this question first can help teams filter out low-impact work before spending too much time considering it.

What is the action we're hoping these people will take?

This is where our discovery-level powerful question can be a great starting point. Again, I recommend Joshua Seiden's *Outcomes over Output* for a great read on how we can think about user behavior-level outcomes in the product development process. (This is a great place to get input from research and design if you have those disciplines in your team or organization!)

What is the value of that action?

This is often the hardest question for us to answer. It's also the one that requires the most creative thinking and the most vigorous desk research. If we want people to make a purchase, what is the average value of that purchase? If we want people to recommend the product, how many new users does each recommendation generally net?

What do we believe is the likelihood that they will complete that action?

Again, this question is nearly impossible to answer, though—as we will discuss later in this chapter—Itamar Gilad's Confidence Meter is a great tool for attaching a numerical score to our confidence that the thing we're building will actually lead to people taking the desired action.

These questions can provide a good starting point for teams to begin mapping their work back to impact-level goals, no matter what steps they've taken to get there. You can find a template for walking your team through these questions at https://mattlemay.com/impact/resources.

Estimating with Unclear or Incomplete Data

Estimating the impact that a small piece of work will have in a complicated and multifaceted world is *really really hard*. Imagine, for example, that a product team at an ecommerce company has been working for a month to make it easier for users to recover a lost password, a task that likely has obvious value to a significant number of users. Suddenly, they are asked by company leadership to describe the "impact" that they believe this work will have on the company's overall revenue targets.

If we have access to the number of our users asking for password resets, we can easily speak to how many people are affected. We can say that we hope that people will reset their password rather than abandoning their cart or even their account. But . . . what is the actual *value* of those people successfully resetting their passwords? And how much would anything we do affect the likelihood of them giving up?

This is where I see many product teams get stuck. They run up against the limits of the first-party data they can access, throw up their hands, and declare the task at hand "impossible" because they simply can't access the information they need. And, in a sense, they're not wrong. **Since this is impact "estimation," there inevitably comes a time where you need to . . . estimate.**

This kind of estimation requires calling upon both first-hand and broadly available knowledge about your product, your business, and your market. The good news is, yours likely isn't the first product team to run up against some common impact estimation challenges. Indeed, the advice I find myself giving product teams when they are faced with the most wicked problems of estimation and attribution? "Start googling it."

So, back to the password reset issue. I began googling the above issues and found the following statistics:

1. "The average U.S. consumer abandons sixteen online purchases a year due to password frustration. This is a staggering number. It means that almost every three weeks, every U.S. citizen is abandoning a purchase online because they forget their password and the process for retrieving it takes too long. And it's a similar story in the UK, with fifteen online purchases a year being left at the checkout by the average consumer." [via iProov]

2. "As for brand reputation, most respondents (53 percent) reported that login problems were a substantial detractor, and an overwhelming majority (85 percent) indicated they look down on a company with identity verification issues." [via Deduce.com]

3. "I run a forum with about 15K active members, roughly one quarter of our daily (unique) visitors use forgotten password processes but of that quarter only half actually complete the reset." [via Serverfault.com]

4. "In its 'Password Usage Study,' HYPR found that 78 percent of full-time workers across the United States and Canada required a reset of a forgotten password in their personal life at some point in the last three months. The rate was slightly lower for work-related resets at 57 percent of respondents." [via Security Intelligence]

Note that a few of these items are deeply anecdotal, and a few of them are published by companies who quite clearly have a vested interest in making this problem space seem *very* important. But, again, we are here to *estimate* impact, not to get it exactly right. And, taken together, these sources begin to present a picture of how a thoughtful product team could put together a more comprehensive case for this work than simply, "I dunno, seems important."

For example, we could imagine a team using that first statistic as a jumping off point for a bit more digging and estimation. Some more targeted googling for "what percentage abandon cart due to password" yields an article from Beyond Identity suggesting that "one in four respondents [to a specific survey] were willing to abandon a cart totaling $100+ if they needed to reset their password to check out." (A company working in identity solutions may not be the most objective source here. But they are also, for that very reason, not likely to *underestimate* the most impact you could reasonably expect from an improved password reset experience.)

From there, this team could find itself in a much better position to estimate the impact of its work. Assuming that $100 isn't terribly far off from the average value of a customer's cart, we could refer back to the questions in our last subchapter to infer that:

> Our customer service team estimates that we have 1,000 password reset requests monthly from users with active carts. We want those users to complete their transactions so that we don't lose revenue. If one in four password-reset requests could lead to a loss of a $100 cart, then we could estimate a maximum impact of $25,000 monthly recoverable revenue from improving the password reset process.

Is a *maximum* reasonable estimation of $25,000 monthly recoverable revenue worth it? That all depends on the team and company's specific goals.

Now let's imagine a team at a *different* company looking at these same statistics. This company is not an ecommerce platform, but rather an ad-supported content website. This could get appreciably more complicated, as ads are served to both logged-in *and* logged-out users, though only logged-in users receive personalized recommendations. A quick Google search for "how much do personalized content recommendations affect ad revenue" yields a McKinsey study (OH NO I MENTIONED MCKINSEY AGAIN) suggesting that "personalization most often drives 10 to 15 percent revenue lift."

Even with that broad statistic as a baseline, a team could begin to run some rough estimates. Going off of our third cited statistic from the prior list, we could estimate that one in four of our total users would do a password reset at some point, and that if only half of them complete the process, we're potentially winding up with one out of eight readers missing out on personalized recommendations. If that results in a 10 percent revenue lift, we could roughly make the following case using our estimation questions:

> We can estimate roughly one in eight of our readers are missing out
> on personalized recommendations due to password reset problems.
> We want those people to successfully recover their password to
> continue receiving personalized articles. If personalization generally
> drives a 10 percent lift in revenue, and our total monthly ad revenue
> from logged-in users is $500,000 of our total $3M ad revenue from
> all users, then we could estimate a maximum impact of $6,250 in
> monthly revenue from improving the password reset process.

This is, obviously, a *very* rough estimate, relying upon questionable statistics from hardly impartial sources. But if you, like me, are surprised that even this generous estimation yields a fairly small number, then you may have just realized why impact estimation is so very important. With practice, working to synthesize proprietary and publicly available information can even become a *fun* exercise. Remember, your job is not to get an exact answer with incomplete information, but rather to get a general sense of what's possible and what's worth pursuing.

Breaking the (R)ICE

Those of you who are familiar with product prioritization frameworks may have spotted some familiar ideas in the impact estimation questions we walked through above. Some of these questions—"How many people will this reach?" "What is the value of the action we are seeking to drive?" and "What is the likelihood of that action being completed?"—map quite well to specific dimensions of common prioritization frameworks such as RICE (reach, impact, certainty, and effort).

RICE is a common twist on the even more common ICE (impact, certainty, and effort) framework used by many product teams to prioritize their work. For a long time, I did not consider frameworks like ICE to be particularly valuable. After all, "impact" can mean a lot of different things to a lot of different people. (As can "effort," for that matter.) But when we define them consistently and deliberately, these three dimensions can prove tremendously helpful in navigating the ever-difficult question, "Is this worth further consideration?"

When I work with product teams that are familiar with the ICE frame-work, I like to frame these three dimensions as follows:

- **IMPACT:** How much could we reasonably expect this work to contribute to our overall goals?
- **CONFIDENCE:** On a scale of one to ten, how likely is it that this work will have the above contribution?
- **EFFORT:** How much time and what kind of resources would we commit to delivering this work?

As we've seen, defining "impact" depends on getting us to a specific estimation that can be expressed in the same unit of measure as our team-and/or company-level goals.

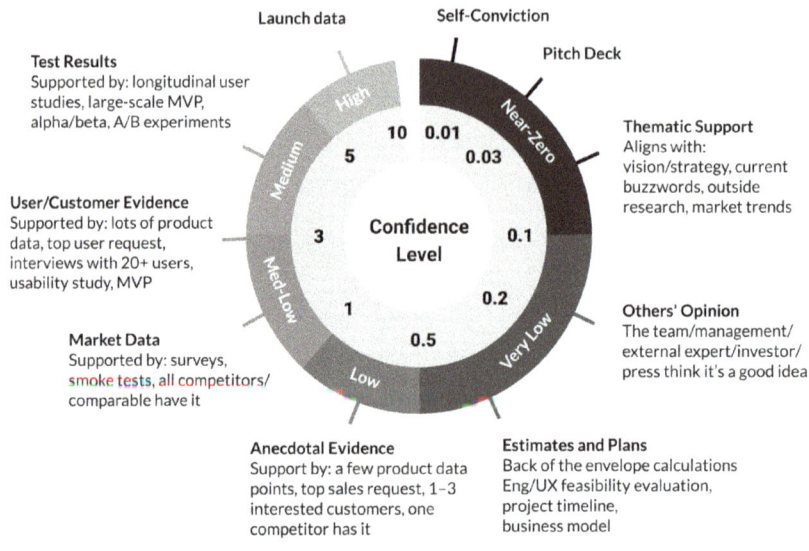

Figure 5-3: Itamar Gilad's Confidence Meter

My favorite approach to scoring confidence is Itamar Gilad's Confidence Meter, a tool that he describes at length in his excellent book *Evidence Guided*. Gilad's emphasis on user evidence and test results reminds us that an impact-first approach to product development does not mean that we ignore our users. It means we have *no choice* but to learn from them directly if we want to deliver work that will be meaningful to our business.

When it comes to effort, I prefer to frame this dimension in terms of both time *and* resources: How long will it take *and* what is the ultimate cost to the business for that time, the time required to coordinate with other teams, and any other required expenses such as new software tools or infrastructure costs? As a starting point, I generally recommend that teams put a rough estimate on the operating cost of their team for a week, so that they can translate the more abstract measure of "time" into actual cost for the business.

The resulting picture can help product teams better navigate the potential trade-offs in their decision-making. If we go back to the story at the beginning of this chapter, for example, the team might conclude that option A would have a potential impact of 1,000 multiproduct users, a confidence score of .8, and would cost the team about four weeks of full-time work which, if estimated at $20,000 a week, would be about $80,000. Option C would have a potential impact of 10,000 multiproduct users, a confidence score of .6 and would cost the team about eight weeks of full-time work plus a fair amount of coordination, which we could round up to $200,000. Is the lower confidence and the extra investment worth a potential 10x return? The answer is not obvious, but at the very least this gives the team the opportunity to discuss these trade-offs openly.

Idea	Impact	Confidence	Effort
Upsell "Super-users"	+1,000 multiproduct users	.8	4 weeks / $80,000
Revamp specific landing page	+100 multiproduct users	*Doesn't matter*	*Doesn't matter*
Add upsell to core onboarding flow	+10k+ multiproduct users	.6	8 weeks inc. commitments from other teams / $200,000

Table 5-1: An ICE Matrix for the product ideas introduced
at the beginning of the chapter

Again, prioritization frameworks like ICE don't take the hard work out of figuring out what to build and why. But with specific impact-level goals in mind, they can help us quickly capture and communicate some of the critical trade-offs at play in prioritization decisions.

Estimation as Conversation Anchor

The quantitative nature of scoring systems like ICE can create an appealing veneer of objectivity and scientific certainty. Just plug a few bits of work into the prioritization machine and BOOM, you'll know exactly what to work on next and why.

The reality is, of course, not so simple. One person's "reasonable best case impact estimate" might be another person's laughable moonshot. One team's "low-effort" feature might require support from a team that is stretched beyond its absolute limit and has no additional capacity to spare. A researcher may have some serious questions for a team that is very confident in their roadmap but has yet to do any . . . you know, research. That's OK. **The goal of impact estimation is to** *structure and facilitate these questions and conversations, not to definitively resolve them.*

Sometimes, these questions and conversations can start to feel like organizational lightning rods. When you start speaking the language of business impact, there is a nonzero chance that your CFO and other assorted executives will suddenly take an acute interest in your work. This sudden interest can be jarring at best and frustrating at worst, and can often be interpreted as critical or downright punitive. But having now encountered these issues from both sides, I can say with confidence that the sudden appearance of previously disinterested executives is a sign that you are communicating the value of your work in clear, important, and broadly accessible terms.

When you speak the language of business impact, **you are putting specific material stakes against work that is usually outside the purview of the folks most accountable for those material stakes.** When those folks come out of the woodwork, they are all but certain to have questions, opinions, and even anxieties to work through. This is a good thing, even if it doesn't always feel like it in the moment. Answer their questions openly and honestly, and be clear about the trade-offs your team has made and why you've made them.

If your team wants to make a difference for the business, these conversations are essential. And if your team proactively addresses business impact, then you have a seat at the table for these conversations and can represent your team's knowledge and expertise.

In Summary

As we move from articulating our goals and ideas to translating them into specific bits of work, it is very easy to incur compounding disconnection and risk. Estimating impact helps us make sure that we are prioritizing and building the things that will actually matter most to the business, no matter how complex and/or convoluted the rest of our processes might be.

Drawing a direct line between day-to-day work and business-level impact is a challenging task for even the most experienced well-resourced product team. But with a little bit of rough estimation, clever googling, and clear-eyed discussion of trade-offs, product teams can keep impact front and center as they prioritize their efforts.

How P&L Responsibility Drove Cross-team Collaboration
Alexander Hipp
Senior Product Manager, Challenger Bank

I was working as a product manager at a challenger bank during a stage of tremendous growth. Up until a certain point, we had been hyperfocused on growing our user base and market share. But as the business matured and investors sought a path to profitability, we had to become more mindful of the money we were bringing in and the money that was going out. To help facilitate this, leadership began slowly introducing the idea that individual product teams would be responsible for their own P&L: for how much money they were bringing in, and how much money they were spending. It was a completely different approach from anything we had done in the past.

The team I was working on had inherited a lot of technical debt, and we weren't thrilled at the prospect of needing to prove the value of our work to the broader business. But once we started looking at the business value of the work we were doing and working closely with our financial strategy team, it actually opened up our thinking quite a bit. Our team was responsible for everything around our premium memberships, and having P&L-level responsibilities left us looking at every single lever we had at our disposal. Not just building new features and writing more code but potentially making subtractions, exploring operational changes, and looking at other opportunities that we hadn't previously considered.

This broader view also changed the kinds of solutions requested by stakeholders, though it didn't change the fact that stakeholders were coming to us requesting solutions! While we had once been given specific features to build, we were now being strongly encouraged to pursue a specific operational change that seemed like an "obvious" way to reduce our expenses. Specifically, we were told to reduce the costs associated with setting up new premium member accounts, knowing that some of these members would abandon or neglect their accounts soon after.

This solution was, in fact, quite obvious. But it also seemed like it was only addressing one small piece of the much broader questions around effectively segmenting and serving our users. If people were not using their accounts, were we funneling them into the right experience? How could we better understand the needs of our users in a way that was actually reducing premium member churn, rather than just mitigating its cost?

With our P&L responsibilities in mind and with the support of our product leadership, we were able to make the case that these broader questions were well worth exploring before we moved ahead with the "obvious" but limited solution that we initially received. To better understand the opportunities at hand, we gathered a group of people representing several relevant teams and roles from across the organization: marketing, operations, UX research, customer support, design, and tech. We agreed to map out the entire user journey together and look for the most impactful opportunities to drive revenue and reduce cost by better understanding and meeting our users' needs every step of the way.

Funnily enough, some of these people had never actually met each other before, even though their work was deeply interconnected as part of a user's experience. **Thankfully, now that teams were responsible for delivering business-level results—not just individual features—there was clear and material value to combining our efforts to uncover the most impactful opportunities.** Rather than leaving our session with a list of features to build, we left with a map of the key metrics we were trying to influence, how they tied in directly to our teams' P&L goals, and the experiments we would run to test and validate a handful of promising interventions, ranging from new features to operational changes to small user interface tweaks.

Needless to say, "We're going to take the time to run some experiments across the user experience," was not as well-received by some of our stakeholders as, "Yes, we're going to go do the thing you suggested right away," would have been. **But because we had mapped out the potential impact we were seeking to drive across the user experience, we were able to make a data-driven case for taking this broader and more holistic approach, rather than debating siloed solutions.** And because we were rolling out smaller experiments rather than fixed releases, we were able to learn and course correct along the way.

We definitely ran into some challenges coordinating across teams as we executed against our plans; that's just the nature of engaging across teams over a long period of time. But the work we were able to deliver wound up doing exactly what we had hoped it would do: reducing costs while also increasing revenue through increased retention and engagement.

CHAPTER 6

Adjusting for Impact

⁣⁣⁣⁣⁣⁣⁣⁣⁣⁣⁣⁣⁣⁣⁣⟶

Keeping impact front and center can allow teams maximum leeway to adjust course as their prior plans and assumptions change. But those plans and assumptions have a nasty habit of pushing our most important goals out of our day-to-day view unless we actively work to keep them visible and relevant. Several years ago, I worked with a growth team at a tech company that had an ambitious, specific yearly goal for growing the company's user base. This was the number that the business had shared in its yearly projections, and the number that shareholders would be looking at to determine whether the company was on the right track. It was, by all accounts, the most important thing for the company to be working towards overall. And certainly the most important thing for a *growth* team to be working towards.

But the broad scale and timing of the company's ambition seemed squarely at odds with the way that this team was being asked to think about and measure their shorter-term progress. In an attempt to better align work across the organization, company leadership was in the process of adopting a quarterly planning cycle using the Objectives and Key Results (OKR) framework. Based on "best practices" for this framework, the team was asked to come up with three to five quarterly goals or "objectives," each of which had three to five measurable "key results" that could be used to track progress.

The process of assembling these objectives and key results took nearly a month to complete. But by the end, the team had about twenty specific key results to measure moving forward. Each key result had an "owner" on the team. Defining their objectives and key results had taken this team a lot of time and effort, and they were keen to get back to their day-to-day work. So things more or less went back to normal, with an agreement to do it all over again during the next quarterly OKR "season."

At the end of the quarter, we set about "scoring" our OKRs: assigning each key result a numerical value from zero to ten representing how much progress we had made. Each key result's "owner" walked through their score and provided their reasoning. And, by and large, the numbers were quite good! There were a few disagreements about whether a particular key result should be scored at, say, a six or a seven. Some team members acknowledged that they had probably aimed too low and could be more ambitious moving forward. But the consensus was that the team had made a *lot* of good progress that quarter.

Towards the end of this meeting, I cautiously shared a question that I had been struggling with for the last several months. "This is great, but do you mind if I ask how much any of this adds up to what we're trying to do by the end of the year? Like . . . **how many new users did we actually add this quarter towards our overall end-of-year goal?**"

After a short pause, a young product marketer on the team chimed in.

"I'm glad you asked that," she said. "To be honest, I'm not sure if *any* of the key results I've been working towards are actually helping us hit our year-end goals."

Another product manager chimed in, "Yeah, I mean, I figured we'd probably just add in those goals for the last quarter of the year, right? Because that's when we'll really need to worry about that number? That's how this is supposed to work, right?"

One by one, everybody turned to the team leader. She let out a deep sigh and stood up.

"OK," she said, "I think we've gotten a bit off-track here." She stood up at the whiteboard and wrote their year-end growth target on the right-hand side.

"This is where we need to get." She then drew a line on the left-hand side of the whiteboard and put a *much smaller number* there.

"This," she said, "is where we are right now. How are we going to get from here to there?"

Immediately, the tone of the conversation shifted. Team members were quick to point out all the reasons why they hadn't made more progress towards their end-of-year goal, from dependencies on other teams to the very fact that they had been asked to go through this time-consuming OKR-setting exercise. But the team leader kept them focused.

"From now on," she said, "I want us to do things a little bit differently. I want you all to be focused on getting us where we need to be. I know it means we're going to have to work more closely with other teams. I know it means we might have to shift around some of our existing plans. But I want to be very clear about this: **If you're even 51 percent sure that a different approach is going to get us 1 percent closer to achieving our goal by the end of the year, I want you to advocate for it.**"

Those words resonated through the team in the months that followed. Ideas and suggestions that would have previously gone unexplored due to cross-team dependencies or deviations from approved roadmaps were brought into the open to be discussed and addressed by the team at large. Each time, the team leader was ready with the same guidance: "If you believe that it's going to get us closer to our goal, then you have my full support." And the next quarter's planning cycle revolved around a single goal for the whole team: getting close enough to our year-end goal that we could be confident that we were on the right track.

I think back to this experience often when I work with teams that struggle to reconcile longer-term impact-level goals with shorter planning and development cycles. At best, impact-level goals give teams the opportunity to constantly readjust their shorter-term objectives and efforts towards a shared, meaningful definition of real-world success. But doing so requires us to continuously work backwards from the impact we seek to drive, rather than starting with our existing plans and hoping it all works out for the best at the end of the quarter or year.

? ONE POWERFUL QUESTION
"What specific shorter-term signals should we see if we are on track to achieving our impact-level goals?"

A high-impact goal several months away can feel distant and far-off to a product team that is navigating shorter-term priorities and pressures. Let's take, for example, our team that is aiming to convert 10,000 single-product users to multiproduct users by the end of the year. If it's only March and we're looking to achieve a specific milestone by the end of December, how do we keep ourselves honest as we move forward?

Again, I find that pulling out a pen and paper can be really useful here. Draw out a timeline, put the milestone on the right-hand side, then work backwards. If it's currently March and we expect to convert 10,000 single-product users by December 31, what do we expect to see before then? When do we anticipate we'll get to 5,000? What else should we be looking at to see if we're on track?

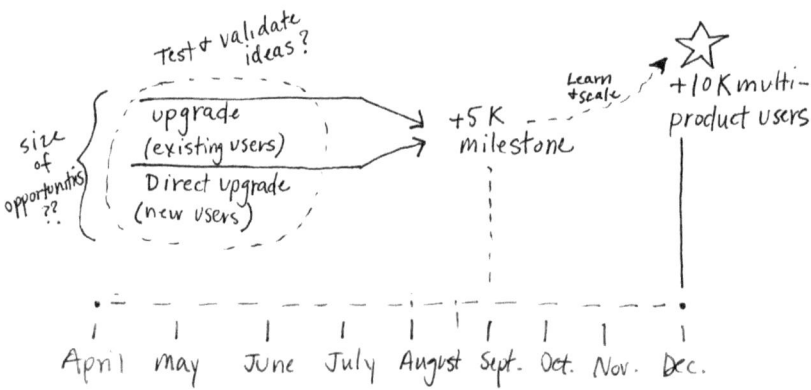

Figure 6-1: A messy hand-drawn diagram for plotting out shorter-term goals and milestones against longer-term, impact-level goals

Looking at the above sketch from our team working towards 10K upgrades, we can see a mix of goals at different altitudes, open questions, and different paths that could be taken. Do we want to upgrade folks who are already extensively using one product? Or is there an opportunity to upgrade newer users during the initial sign-up flow? How big is each of these opportunities? And should we aim to have 5K of our 10K total users upgraded? **It is through this process of backing out from outwardly simple impact-level milestones that we often get out ahead of the complexities and uncertainties that we will need to address along the way.**

I particularly like this approach because it helps teams move past an abstract conversation about "leading" and "lagging" indicators, and towards a conversation about what specifically they will measure, when, and why. At organizations with long sales cycles (such as many enterprise B2B companies), critical metrics like revenue may very well "lag" well behind a product team's work, prompting them to seek out other ways to measure their success along the way. At organizations with a more rapid and trans-actional value exchange (such as many B2C and ecommerce companies), revenue and user growth might be easier to project and measure directly at regular, shorter-term milestones.

In the case of the team we met at the beginning of this chapter, new users were signing up and old users churning on a daily basis, meaning that we could map out our intermediate milestones using the same metric as our ultimate goal. Slotting in planned launches and initiatives throughout the year—and estimating the impact of those launches and initiatives—helped us get a more comprehensive picture of where and when we would need to explore new avenues for growth. And, most important, **mapping out our intermediate goals and milestones meant that we couldn't just sit back and hope for the best. We had to actually *deliver* things on a regular cadence if we wanted to make any real, measurable progress.**[1]

Establishing a Cadence for Changing Course

As we discussed in chapter 2, starting with impact gives teams the maximum flexibility to evaluate any and all approaches that might best contribute to a successful business. When teams are accountable for high-impact goals,

---

[1] The artifact that we create when we work backwards from our impact-level goals could be described as an "outcome-based roadmap." It maps out the goals we are working towards which can, in turn, help us better scope and prioritize the work we plan to deliver. Pretty cool!

they are also accountable for changing course as needed if their current approach is not helping them deliver against those goals.

But that, of course, only works if teams actually do the work of changing course.

When I was researching my book *Agile for Everybody*, my approach to changing course . . . well, changed course. Dramatically. Previously, I had tried to make room for flexibility by discouraging long-term planning. The fewer long-term plans we had to follow, I reasoned, the more flexible we would be. But in an interview that truly shifted my perspective, Agile practitioner and advocate Kathryn Kuhn shared the following insight:

> We can't always get an organization to stop doing years-long planning cycles. But we can get them to add a quarterly business review. It's pretty simple—you start reflecting on how you did last quarter, then you bring in additional information that you want people to know. Maybe there's a new Gartner study about how the market is changing, new regulatory requirements, new initiatives from company leaders, or new information about our customers. Bring that information into the room, describe what you've learned since the last time you discussed your plans, and then look ahead. Can you look at some of your initiatives and, based on what you now know, declare them "done enough?" If you're 85 percent of the way toward achieving your goal with one initiative, can you strategically shift your resources to another initiative where you're only 20 percent done?

In other words, the key to adaptability is not less planning, it is *more* regular, choreographed, and *foreseen* planning. If changing course fundamentally *undermines* a plan, it is unlikely to happen—especially if some very important people signed off on that plan. If changing course *is part of a plan*, it gives us much more flexibility to ask and answer big, challenging questions. Product teams that only plan for small changes only make small changes. Product teams that actually plan for big changes—that are willing to periodically look at the work they're doing and ask, "Is this still going to get us where we need to be?"—are much more likely to make the changes they need to *get* where they need to be.

For that reason, I recommend **establishing a regular cadence to evaluate progress and ask those big, challenging questions about whether the approach you're taking is actually helping you deliver success for the business at large.**

Here are some questions you can ask to make sure you're keeping impact in mind while navigating the day-to-day demands of product work:

Have the mission-critical milestones we're working towards changed?

Sometimes, the broader business context shifts in a way that requires us to fundamentally reevaluate what we're working towards and why. (For example, if a business suddenly announces that it is pursuing a substantial acquisition.) Discussing these macrolevel changes regularly can help ensure that an individual team doesn't lose its connection with the overall success of the business.

Do we believe that our team's specific goals still represent the best possible way to achieve our desired impact?

Even for teams who have set high-impact, high-specificity goals, there is always a chance that those goals could change. For example, the team we've followed driving multiproduct usage could, in the course of executing their work, realize that they would be able to *better* contribute to the company's overall revenue goal by bringing in new single-product users, or by looking at new ways to package and commercialize different products for different user segments.

How confident are we that we will achieve our desired impact if we stay on our current course?

This is, perhaps, the most important question to ask, since it opens up a space to address concerns that might be present in degrees, not absolutes. I often ask each member of the team to rate their confidence individually from one to ten, then to compare notes to identify uncertainties and misalignments. Through this process, we are able to bring uncomfortable questions into the open that might have otherwise gone unasked and unaddressed.

Is there any work we're currently doing or currently have planned that we *don't* believe is helping us achieve our desired impact?

As product teams go about their day-to-day business, there is almost always pressure for them to take on work that might not be aligned with their most important goals. Acknowledging this reality can help teams realign and regain focus. It can also help reallocate resources to the things that are actually going to deliver meaningful results for the business.

How has the broader market changed since our last check-in, and what does it mean for our users?

The world changes quickly, and it is almost certain that the broader market—competitive landscapes, needs and goals of users, the macroeconomic environment—has changed since the last time you stopped to take stock of it. Making time to regularly reflect on these changes can help your team stay grounded in the fast-moving conditions of the world around you, rather than using them as an excuse later on.

You can find a template for walking your team through these questions at https://mattlemay.com/impact/resources. Regardless of the specific questions you ask and conversations you facilitate, remember that **meaningful changes in your overall approach are unlikely to happen unless you make them *part* of your plan.** Having some intermediate milestones, objectives, and questions mapped out can help structure the way your team navigates these changes, and create shorter-term urgency around longer-term goals.

When It's Not Going Well

Because impact-level goals are outside a team's immediate control, it can be very easy to point fingers in any and all directions when these goals seem out of reach. Sales and marketing functions, company leaders, well-funded competitors, cultural and economic shifts, and *many many more things* can become convenient scapegoats when a team begins to feel like its high-impact, high-specificity goals are slipping away.

As my friends who ran events businesses during the 2020 lockdown can attest, there absolutely *are* times when broader circumstances limit what's possible for a team or organization to achieve. And as the next sidebar story will illustrate, there are times when new priorities force us to readjust our ambitions and timelines. (Again, I really hope you're reading those sidebars!)

There are also times when, quite frankly, something that we hoped would deliver a lot of impact seems more and more likely to land with a deafening thud. We've all been there, and it's never much fun, no matter how much our companies insist that they foster a culture of "failing fast."

The good news for impact-first teams is that having high-impact, high-specificity goals can make it easier to adjust course *into* work that matters most, rather than *away from* work that has ostensibly "failed." John Cutler put it perfectly in a recent newsletter when he provided his own take on Goodhart's Law, which we discussed in chapter 2:

In environments with high psychological safety, trust, and an appreciation for complex sociotechnical systems, when a measure becomes a target, it can remain a good measure because missing the target is treated as a valuable signal for continuous improvement rather than failure.

In other words, targets can be tremendously useful in environments where they're used as tools and guideposts, rather than bludgeons.

I'd go one step further and suggest that **teams that proactively and honestly track progress against impact-level goals are helping to model what it looks like to value "psychological safety, trust, and an appreciation for complex sociotechnical systems" in practice.** They are able to direct conversations away from "whose fault is this?" and towards "what have we learned, why does it matter, and how can we readjust our work and resources accordingly?"

Setting shorter, planned cadences for adjusting course—and knowing what signals to look for and which actions to take—can help shift the conversation from, "Oops, we missed our targets" to, "It looks like we need to change our approach in order to have a better chance of delivering success for the business, here's why, and here's what we're doing about it." For teams that are in the habit of regular retrospectives and practicing continuous learning, it also affords the opportunity to ask, "What assumptions did we make at the outset that turned out to be untrue, and what can we do to avoid such assumptions moving forward?"

When It *Is* Going Well

For teams that are new to tracking their progress against business-level impact, a clear path to success can be scarier than an uphill and obstructed one. I recently worked with a team that had tied its success to the company's overall growth goals and, a few months in, recognized that the company itself was growing at a rate that exceeded their ambitious targets. But the team's product manager still seemed quite uneasy. During a coaching session, she confided in me that the company's top-line success had actually left her feeling *more* exposed as a member of her immediate team: "The company is doing really well, but I'm having a hard time explaining why and how *we* have contributed to that success."

And therein lies another challenge for impact-first product teams: Just as failure to achieve broader impact-level goals can always be attributed elsewhere, so too can success. Just as low-performing teams elsewhere in the organization can be easily scapegoated, so too can high-performing teams monopolize attention and praise. And just as negative market conditions can derail your team's progress, so too can positive market conditions carry your team to success that feels tenuous, qualified, and unattributable.

This sense of ambiguous dread can drag down even the most outwardly high-performing product teams. I have seen many such teams inadvertently downplay the broader success of the organization if they don't feel *uniquely* responsible for that success. I have also seen several teams subtly downplay the contributions of *other* teams in the hopes of making their own contributions seem greater by comparison. Both of these approaches make it harder to maximize a company's success when things are going well, and all but impossible for a company to work together effectively if and when things start going poorly.

Truly impact-first product teams shift the conversation from "how can we isolate and track *only* our contribution?" to "how can we understand

and celebrate the ways in which *every* team worked together to deliver success?" If, for example, an impact-first product team sees sales and marketing teams driving an outsized contribution, they *reach out* to folks in sales and marketing to understand what those teams are doing well and how they can best collaborate moving forward. When multiple teams can learn and celebrate together, these teams are more likely to work together when things aren't going as well . . . and when it's harder to tell how things are going at all.

When It's . . . Complicated

For most teams, most of the time, things are rarely as simple as, "We will never achieve our goals and need to change course" or, "We will definitely achieve our goals and need to celebrate our wins together." When I ask folks how their teams are progressing towards their goals, I'm much more likely to hear something like, "Um, you know, I think we're doing OK?" or "It's tough to really know how we're doing with all that's happening?" Certainty is hard to come by, and upward vocal inflection abounds.

For some teams, the easiest way to manage this uncertainty is to retroactively disconnect the work they're doing from the desired impact of that work. If we aren't sure if a given piece of work is on track to do much for the business, we can always just mark it as "in progress" and continue dutifully marching towards its completion. Or, if our team tracks the potential success of its work using a traffic light system, we can simply assign a "yellow light" status to nearly everything that isn't an obvious success or a near-comical failure.

Many of my recent conversations with product leaders have revolved around the dangers of ambiguous-status categories such as "yellow light" or "in progress." These conversations have compelled me to reflect more critically on my *own* tendency to create a kind of "eh, don't worry about it" bucket of work when tracking real-world impact seems too complicated or challenging.

This bucket has often taken the form of the "yellow light" label against which I have been recently, repeatedly warned. But in other cases, it has consisted of work marked "must complete" or "quick win" or "just need to get done" or "business as usual." And every time I have created such a "don't track impact because *reasons*" bucket, that bucket has slowly grown to encompass nearly all of my team's work. **If you give your team the opportunity to flag certain bits of work as beyond impact-level scrutiny, it is almost inevitable that *all* of your work will begin to creep away from impact-level scrutiny.**

So what do you do when your team's work . . . may or may not be on track to deliver impact? Again, there is no one-size-fits-all answer here. But the question at the beginning of this chapter provides a helpful start: What are some earlier signals we can look for, faster experiments we can run, and other shorter-term steps we can take to better understand whether our current course is likely to be successful? Broadly speaking, **I recommend taking all the tools and techniques we discussed in the last chapter on impact estimation—especially Itamar Gilad's confidence meter—and treating them as ongoing activities rather than one-time exercises.** Understanding and estimating business impact is not a single point in the process of product development, but rather an ever-present discipline and mindset we bring to our work as impact-first teams.

(In the interest of avoiding dogma and, therefore, hypocrisy, I do feel compelled to note here that I have worked with a few teams that have successfully implemented *deliberately limited* buckets of "just needs to get done" work. If you really *do* need to create a category of work that isn't being tracked against impact, please, please *constrain* that category of work. Intentionally setting aside 10 to 20 percent of your team's capacity for work with ambiguous impact puts you in a much better position than letting 100 percent of your team's work drift into a state of liminal "I dunno"-ness.)

When There's Something "Really Important" that You Just "HAVE" to Do

If I could summarize my last book *Product Management in Practice* in a single sentence, it would be: "The people at your organization telling you to 'just build something' might know things that you and your team do not, and you should approach them with curiosity rather than dismissing them out of hand."

For an impact-first product team, being told by an important person to "just go build something" is *not* proof that they work for a "feature factory" and that all their efforts to focus on business impact have been fruitless. Instead, it is a way to open up a conversation that might reveal previously unknown information. Have broader conditions changed? Is the company measuring success in a different way, even if it hasn't officially been announced or communicated? Are other teams working towards goals that might be in conflict with my team's goals?

And that's the thing: In an organization of any appreciable size, *everybody* has incomplete information. Your company's CFO might have phenomenal insights into the overall financial health of the company, but might be missing some key information about the particular customer problem your team is working to solve. **An "emergency" or "urgent" request to do something—or the insistence that something is "just obviously the right thing to do"—can be an important jumping-off point for getting a more comprehensive picture of the fast-changing and complex circumstances at hand**. But only if you treat it as such.

Thankfully, an impact-first product team is well-versed in drawing connections between *output* and *impact*. And an impact-first product team is open and receptive to changing their output if that shift is likely to deliver more impact for the business.

Broadly speaking, when approached with urgent and politically loaded requests, I've found it most helpful to respond with neither a "yes" nor a "no," but an inquisitive and open: "Wow, thanks so much for bringing this to me. My team is currently working on [output] because we believe it will deliver [impact]. Could you help me understand why you're bringing this to me? I'd love to know more about how it can help us achieve our goals."

This openness and curiosity allows product teams to shift away from best practice based-dogma and towards working within the commercial realities of their business. Commercial realities which, as we discussed earlier, might not have been an issue for the companies from which those "best practices" emerged.

For example, I recently worked with a small B2B software company whose product manager was incredibly frustrated that their CEO had been telling him to build things that only *one* customer was requesting. Sure enough, I have read countless articles on LinkedIn describing why "building for one customer" is a bad, *bad* idea. But after a few conversations among the broader leadership team, it became clear that this one customer actually represented nearly half of the company's monthly recurring revenue, and the risk of losing them far outweighed the benefit of building any of the other things the product team had scoped. The conversations that followed from this understanding helped the product team actually understand this customer's needs, concerns, and how best to keep them happy while also considering ways to grow and diversify their business overall.

To be clear, these conversations can be tricky. But when they play out at the level of business impact, it is much easier for product teams to acknowledge "we need to build this because our company's existential success is on the line" as opposed to "we need to build this *because an annoying executive told me to, even though it's not the right way to do things.*" From this less defensive position, product teams are free to do better work without getting wound up and stressed out about deviating from some abstract ideal of how things are supposed to be done.

In Summary

Keeping business-level impact front and center as we go through the day-to-day work of our product teams can be challenging, especially when the impact we seek to drive is hard to measure in shorter increments. But by working backwards from our overall goals and being clear-eyed and candid about the progress we hope to see—and the decisions we might need to make if we *aren't* seeing that progress—we can keep impact top-of-mind as we go about the complicated and challenging work of building and delivering products.

Revenue Forecasting to Facilitate Communication and Drive Impact

Zak Jallaly
Product Manager, Dating App

I was working as a product manager for a dating app that was exploring some new features that might fundamentally shift its unique positioning in the market. There was a fair amount of anxiety around these features, and uncertainty about what the far-reaching ramifications might be throughout the app and company as we moved forward.

I come from a fairly commercial background, so as we began developing these features, I wanted to keep monetization top of mind from the outset. **After all, big-picture trade-offs are even harder to navigate if you don't have a general understanding of how much value for the business is on the line.** We ran a few experiments, tried a few things, gathered some qualitative feedback, and felt pretty confident that we had something that could be effectively monetized and still improve the overall user experience.

This is where things get really interesting. Going into the next year, the organization gave my team a pretty tough goal to clear: we were responsible for generating an amount of revenue that required us to 4X the annual run rate of a multimillion-dollar product line. I'm not sure if they could have given us this kind of a goal if not for the work we had done to explore and communicate our work's commercial potential and the feature's market fit with our user base. But here we were, facing down twelve months of big question marks with an ambitious and challenging impact-level goal at the end.

We knew we weren't going to get there without a plan. So we sat down as a team and said, "What do we need to do to get there by the end of the year? What will we need to accomplish along the way?"

We started with weekly and monthly forecasting, building on the real-world data we had from our earlier experiments with the feature. On a weekly basis, how many dollars do we need to be making? Then, what are the levers we have to play around with towards achieving those revenue goals? Some of the levers we landed on were, for example, users engaging with our feature and then going on to pay for it, users paying for our feature directly, how much they're paying—pretty basic awareness, adoption, conversion metrics that can all be mapped back to specific dollar amounts. We looked across the year holistically and made a plan that geared into the broader plans of the organization: "In December, we can expand this to other markets, so adoption will go up overall. In May, we can do a bunch of initiatives to increase user adoption. When should we plan to look into pricing adjustments that would increase value per user?"

The forecast was valuable for a handful of reasons. First, it kept us on track to adjust course and make important decisions along the way. On a monthly basis, do we need to increase our efforts? Reduce our efforts? Move from one lever to another? Second, it gave us cover to work more closely with other teams and stakeholders—we were a matrixed organization, there was no way we would be able to achieve such an ambitious revenue goal without working in close partnership with marketing teams, for example. Third, it created a lot more trust between us and executive stakeholders, as we could speak a common language rooted in the success of the business.

All of these reasons proved particularly important as we dealt with the twists and turns of working in a high-usage and high-stakes product organization. At the beginning of the year, we encountered a few serious and unforeseen challenges that took our attention away from planned work. **But because we had taken the time to map out our forecasts for the year, we were able to go back to the organization at large and adjust both our year-end goals and our intermediate plans along the way.** I'm very happy to say that we were able to hit those adjusted targets by the end of the year, and even happier to say that we were able to get there despite the year not shaping up exactly how we had imagined.

CHAPTER 7

The Journey of Impact-first Teams

Though it involves its share of numbers and mathematical estimates, becoming an impact-first team is more of an art than a science. Product work is intrinsically communicative and political work. Drawing a line between that work and the existential success of the business can bring uncertainty and assumptions to the table to be addressed together, but it can't definitively resolve them through mathematical magic. There's a reason each of the preceding chapters revolves around a question, not a directive: **the conversations you have together will be far and away the most important part of your journey to becoming an impact-first team.**

Every team's needs, goals, constraints, and experiences are appreciably different. The stories in this book, and the steps and exercises contained therein, represent snapshots of how specific teams have done the challenging work of bringing business impact into the heart of what they do.

To extrapolate out a single, one-size-fits-all "process" from these stories and exercises would be to underestimate the differences between teams. Similarly, to suggest that there is a simple "formula" for determining or working with impact-level goals would be to underestimate the broad sets of circumstances in which teams and companies can find themselves. The work of becoming an impact-first team is *your* work, to be done in your particular context with your particular teammates.

In this chapter, I've put together some of the specific questions and broad steps I've helped teams navigate through my consulting work. Again, some of these will likely resonate with you more than others, and that's OK. Rather than treating the following sections like a one-size-fits-all to-do list, I recommend reading through quickly and looking for the questions that you believe are most likely to kick up interesting and valuable conversations among your team, and start there. If you'd like a copy of these questions to share with your team directly, you can find one at https://mattlemay.com/impact/resources.

1. Bringing impact into the open

Our powerful question from chapter 1: "If you were in charge of the company, would you fully fund this team?"

Other conversation starters to bring to your team:
- If an executive cornered you and asked you why your team's work is important for the business, how would you answer?
- If this team and its work disappeared tomorrow, what impact would it have on the business?
- Is there another team whose work you would consider particularly impactful? Why? And how can you learn from them and their approach?

After working through these questions, your team may have:

The sense of cautious resolve that comes from having difficult conversations about your team's impact on your own terms, rather than waiting until you are forced to have them on *someone else*'s terms.

If you get stuck or can't get there, you could try:
- Having candid 1:1 conversations with your teammates where you openly share your concerns and questions
- Getting into the habit of regularly retrospecting with your team to bring up these questions in context
- Skipping ahead! Not every team is ready to have these conversations right away.

2. Understanding what impact means to the business

Our powerful question from chapter 2: "What are the measurable conditions that must be met for our business to be successful at a specific point in the future?"

Other conversation starters to bring to your team:

- Do we know what the next major milestone is for the business, such as raising a round of funding or releasing a quarterly earnings report?
- Are there any business-level goals or metrics that make us anxious when we're *not* contributing to them?
- Do we understand how our business model works and how the company funds and sustains itself?

After working through these questions, your team may have:

At least one goal that you know is important to the company, and an understanding of why it's important. Try using the format:

The company's existential success hinges on achieving [milestone] by [date] because [reason].

For example:

> The company's existential success hinges on achieving [profitability] by [the end of the year] because [it has promised its investors that it will do so].

> The company's existential success hinges on achieving [10,000 monthly active users] by [the end of the quarter] because [it is running low on cash and it will be very hard to raise the next round of investment without a sizable user base].

If you get stuck or can't get there, you could try:

- Looking at company town hall meetings, public reports, and other documents. Remember you don't need to have a "perfect" answer, just a place to start.

- Asking folks in leadership positions what matters most to them and what keeps them up at night
- Starting with team-level goals and socializing them with the broader organization. In the absence of clear company-level goals, sharing team-level goals can be a great way to initiate a conversation around what the organization at large needs to achieve and why.

3. Setting and confirming impactful team goals

Our powerful question from chapter 3: "What measurable contributions will this team make to to be considered a successful part of the business at a specific point in the future?"

Other conversation starters to bring to your team:

- What would this team need to accomplish for us to all feel confident about our contributions to the business by the end of the quarter or year?
- Are our team-level goals both high-impact (no more than one step away from company goals) and high-specificity (attached to specific targets and timelines)?
- Which stakeholders in the organization will be evaluating our team's success, and what matters most to them? (e.g., Department heads, other product managers and teams.)

After working through these questions, your team may have:

A set of high-impact, high-specificity team goals. Try using the format: Our goal is to achieve [measurable result] by [date] because [reason]. For example:

> Our goal is to [convert 10,000 of our single-product users to multiproduct users] by [the end of the year] because [the customer lifetime value of multiproduct users is greater, and this will contribute meaningfully to the company's revenue goals].

> Our goal is to [onboard 1,000 of our existing users to the new version of our platform] by [the end of the quarter] because [the business has devoted a lot of resources to building this and shareholders expect to see material progress].

If you get stuck or can't get there, you could try:

- Evaluating multiple orders of magnitude to find a "good enough" target. (Remember our checklist from chapter 3!)
- Working with other teams in the organization to see how your goals might be interrelated and how you can work together to maximize your impact
- Putting together a few different options for team-level goals and evaluating them with your team leadership to see what resonates the most and what trade-offs are at play

4. Impact through the middle

Our powerful question from chapter 4: "Who needs to do what for us to achieve our impact-level goals—and why would they do it?"

Other conversation starters to bring to your team:

- Do we have a clear sense of who our users are, what they care about, and how they exchange value with us?
- What set or type of users should we focus on to achieve our specific team-level goals?
- Who else in the organization (research teams, customer support agents, etc.) has first-hand contact with our users? What could we learn from them?

After working through these questions, your team may have:

Some initial questions or focus areas you can bring to the user researchers in your organization or on your team. Or that you can explore yourselves if your organization does not have dedicated user researchers.

If you get stuck or can't get there, you could try:

- Involving user researchers in your organization as early as possible
- Reading through existing insights, segmentation reports, and other artifacts that could help you understand your user base
- Looking up information about the users of your general type of product to understand macrolevel trends and opportunities

5. Estimating impact

Our powerful question from chapter 5: "What is the most impact we could reasonably expect from a given piece of work?"

Other conversation starters to bring to your team:

- If our team were to suddenly complete all of our planned work, what impact would it have on the business overall?
- Are there any subtractions, deletions, copy changes, or other things that are not *new features* that might help us achieve our team-level goals?
- If none of the things our team plans to build seem poised to achieve meaningful impact, is it worth pursuing a change in our team's purview or realigning our resources with another team?

After working through these questions, your team may have:

A way of understanding and quantifying the "impact" of your team's work that is rooted in specific and meaningful goals, as opposed to arbitrary scores or, you know, *vibes.*

If you get stuck or can't get there, you could try:

- Googling! Remember, a lot of wicked estimation problems have already been worked through by lots of other teams.
- Talking to folks in your organization who have approached similar work or sought to achieve similar goals
- Reevaluating your team's "mission" or purview to see if it is broadly in line with your impact-level goals

6. Adjusting for Impact

Our powerful question from chapter 6: "What specific shorter-term signals should we see if we are on track to achieving our impact-level goals?"

Other conversation starters to bring to your team:

- What are some steps we can plan to take if we are *not* on track to achieving our impact-level goals?
- How do we adjust our planning practices and cadences to allow maximum flexibility for us to achieve our impact-level goals?
- What is the *last* minute at which we could shift our broad plans if we're not on track to achieve our impact-level goals?

After working through these questions, your team may have:

A set of shorter-term cadences and milestones rooted in the realities of what your team and business need to achieve.

If you get stuck or can't get there, you could try:

- Incorporating your impact-level goals into existing planning practices and cadences to see what questions arise
- Retrospecting on your team's last year or quarter to see when and where it would have been helpful to measure progress and adjust course
- Attaching impact-level goals and progress to a past roadmap to see how outputs and impact are connected for your particular team or organization

Every team's impact-first journey is different, and few are tidy or linear. But these questions and approaches can be helpful for teams looking to break out of the Low-impact Death Spiral, do meaningful work for the business, and work towards demonstrable and broadly understandable success.

Letting Your Job Be Your Job

▪▪▪▪▪▪▶

Managing business-level expectations might be the most challenging part of product work. Product teams often bear the brunt of business outcomes that seem beyond their purview or pay grade, which can leave folks in the incredibly stressful position of feeling responsible for things well beyond their control. My biggest hope in writing this book is that it will show us, as product people, a way to take the conversations we dread the most and put them at the heart of our work in a reasonable, clear-eyed, and collaborative way.

Putting impact first gives us a unifying language to align the work we do across functions and departments. It gives us a way to align our interests with the success of the businesses we serve. It gives us a way to create urgency around understanding the needs of our customers.

Most of all—and quite counterintuitively—I hope it gives us a way to remember that, for better or worse, our job is just a job. We are not on a sacred mission to get people to build products the "right way." We are not fearless warriors fighting for "the customer" at the expense of "the business." Whether we are product managers, designers, engineers, program managers, business analysts, or some new title that I'm somehow going to have to come up with an opinion about in the next month, our job is to contribute as best we can to building a successful business. No more, no less.

I truly, *truly* wish I had understood this earlier in my career. I wasted a whole lot of energy fighting Quixotic battles against . . . um . . . *the commercial realities of the companies that were paying my rent.* If I had redirected even a tiny bit of that energy to better *understanding* the commercial realities of those companies, I would have been a much better product manager. And probably a much happier person too.

I hope that the stories in this book point to a clearer, calmer, less embattled future on the other side of some important conversations about our business, its goals, and our contributions. And I hope that the questions in this book help you and your team approach those conversations with candor and purpose. You got this!

Work With Me

━━━━▶

As you may have noticed, most of the stories in this book come from my work with product teams and organizations. I've worked globally with companies ranging from startups to Fortune 500 enterprises, and have yet to find one that couldn't use some help maximizing the impact of their product managers, teams, and organizations.

I love doing this work, and I take pride in doing it in a way that is focused on relationships and results. I don't make a lot of PowerPoint decks or use a lot of heavy frameworks. I'm not looking to build a massive consultancy at scale. When you work with me, you're working with *me*.

Here are some of the specific ways I engage with companies:

Product Team Coaching

Helping product teams maximize impact and minimize busywork is one of the great joys of my professional life. As we discussed in chapter 1, every product team has the potential to break the Low-impact Death Spiral. But they are far more likely to unlock this potential with some expert help and guidance.

Product Leadership Consulting and Advising

Strong product leadership can be a powerful catalyst for developing and nurturing impact-first teams. I love working with product leaders to cocreate an impact-first approach to managing, inspiring, evaluating, and incentivizing their teams.

Goal Setting and Alignment Sessions

Setting and aligning team- and company-level goals does not need to be an onerous or exhausting process. And it certainly does not need to be a yearly or quarterly exercise that the organization slogs through only to return to business as usual. Show me your team- and company-level goals and I'll show you how to make them clearer and more impactful.

Workshops and Training

I offer a variety of half-day, one-day, and multiday workshops and training programs to introduce the concepts in this book to product teams and organizations that want to put impact first. These workshops can be customized to suit the particular needs, goals, and scale of your organization.

You can find more information about my consulting services at http://mattlemay.com/consulting or by contacting me at consulting@mattlemay.com.

Acknowledgments

Thank you mom for being a true writer.

Thank you my wonderful sister Saielle DaSilva for making this book a million times better.

Thank you Sally McGraw for the guidance and patience.

Thank you Todd Goldstein for keeping it classy.

Thank you Christi Williford for making it make sense.

Thank you Carrie Tian for showing me what book I was writing.

Thank you Petra Wille for strategy and tactics.

Thank you Martin Eriksson for thoughtful descoping.

Thank you O'Reilly Media for giving me a platform.

Thank you everyone who has trusted me to work, build, and learn with their teams.

Thank you early readers for the feedback and encouragement.

Thank you European Product Land Community for letting me sneak in.

Thank you Frances for seeing me and believing in me.

About the Author

Matt LeMay is a product leader, consultant, and author. He empowers product managers, teams, and organizations to maximize their business impact by streamlining processes, simplifying strategies, and focusing on the work that matters most.

Matt's decade-plus career in product has included acquisitions by Google and Intuit, strategic advising to Spotify, and building out product teams for early-stage startups that are now valued in the hundreds of millions of dollars. His books *Agile for Everybody* (O'Reilly Media, 2018) and *Product Management in Practice* (Second Edition O'Reilly Media, 2022) have been translated into more than six languages, and provide actionable guidance to individuals and teams across countries, functions, and industries.

Matt is the creator of the One Page / One Hour Pledge, a commitment to minimize busywork and maximize collaboration that has been adopted by more than 100 individuals and teams at Amazon, Walmart, Adobe, Disney, and more. His work with tech ethnographer Tricia Wang has been included in Google's official Design Sprint toolkit.

Previously, Matt worked as Senior Product Manager at music startup Songza (acquired by Google), and Head of Consumer Product at Bitly. Matt is also a musician, recording engineer, and the author of a book about singer-songwriter Elliott Smith. He lives in London, England.

Index

〰〰〰➡